T0362102

L-System Fractals

This is volume 209 in
MATHEMATICS IN SCIENCE AND ENGINEERING
Edited by C.K. Chui, *Stanford University*

A list of recent titles in this series appears at the end of this volume.

L-System Fractals

J. Mishra
DEPARTMENT OF COMPUTER SCIENCE AND APPLICATIONS
COLLEGE OF ENGINEERING AND TECHNOLOGY
BHUBANESWAR, ORISSA
INDIA

S.N. Mishra
DEPARTMENT OF COMPUTER SCIENCE AND APPLICATIONS
INDIRA GANDHI INSTITUTE OF TECHNOLOGY
SARANG, DHENKANAL, ORISSA
INDIA

ELSEVIER
Amsterdam – Boston – Heidelberg – London – New York – Oxford
Paris – San Diego – San Francisco – Singapore – Sydney – Tokyo

Elsevier
Radarweg 29, PO Box 211, 1000 AE Amsterdam, The Netherlands
The Boulevard, Langford Lane, Kidlington, Oxford OX5 1GB, UK

First edition 2007

Notice
No responsibility is assumed by the publisher for any injury and/or damage to persons
or property as a matter of products liability, negligence or otherwise, or from any use
or operation of any methods, products, instructions or ideas contained in the material
herein. Because of rapid advances in the medical sciences, in particular, independent
verification of diagnoses and drug dosages should be made

Library of Congress Cataloging-in-Publication Data
A catalog record for this book is available from the Library of Congress

British Library Cataloguing in Publication Data
A catalogue record for this book is available from the British Library

ISBN-13:   978-0-444-52832-2
ISBN-10:   0-444-52832-6
ISSN: 0076-5392

For information on all Elsevier publications
visit our website at books.elsevier.com

Transferred to digital printing 2007

Printed and bound by CPI Antony Rowe, Eastbourne

Preface

Many objects in nature, like trees, mountains and seashells, the nervous system of human beings, etc, have a property called self-similarity. Sometimes this property is very pronounced. Other natural phenomena exhibit this property to a lesser degree. Recognition of these natural shapes has proven to be a challenging problem in computer vision. The members of a class of natural objects are not identical to each other. They are similar, have similar features, but are not exactly the same. Most existing techniques have not succeeded in effectively recognizing these objects. One of the main reasons is that the models used to represent them are themselves inadequate.

There are many methods for generating fractals. Out of these methods, Iterated Function Systems (IFS) and Lindenmayer Systems (L-Systems) have gained the maximum popularity among modelers for generating fractals for natural objects. There are several characteristics which are associated with fractals; e.g., self-similarity, chaos, dimension, etc. However, besides some known close relations between iterated function systems and Lindenmayer systems, many other relationships are still not exhausted. Almost every fractal figure can be generated using the L-System concept. This book makes an attempt to explore a few facts about the generation of fractals using the same.

This book is based on research findings of our PhD work on characteristics of fractals using the L-System concept. It strives to motivate intuition about simulation and modeling of natural fractal figures so as to present them in the concise codes of L-Systems. It is an attempt to give a comprehensive and state-of-art treatment of important aspects of fractals to be generated by IFS and L-System concepts. There are many references to future directions as well as present work done so far in order to help researchers to think about the concept.

The book can serve as a definite reference for simulation practioners and researchers in the area of fractals and computer graphics. We have also made major efforts to link the subjects to relevant research literature, both in print and on the web, so as to keep this material up to date.

Prerequisites for understanding the book are a knowledge of basic concepts of fractals and regular grammar used in compiler designing.

The eight chapters in this book cover:

- an introduction to fractals, computer graphics and L-systems
- a survey of fractals and L-system concepts
- generation of fractal figures using IFS
- generation of a new class of hybrid fractals
- generation of L-system strings using ramification of matrix
- practical modeling and rendering of fractals
- fractal dimension calculation using traditional and new concepts
- research directions in the field of L-system concepts

The Appendix of our book has three parts. Appendix-A gives the L-system codes for some converged fractal figures in Chapter 4. A few novel fractals are generated by using combinations of Koch curve and its variations with that of 1 to n series, whose L-system codes are written in Appendix-B. A few L-system codes are given in Appendix-C for the creation of fractals using the concept of Ramification Matrix.

In the colour section of the book there are a few colour plates, which exemplify the potential of the methods, used for generating fractal figures in Chapter 3 and some figures mentioned in chapter 8 of a research work found in the literature. Numeric references have been given in various chapters within square brackets in order to link the contents easily mentioned in the bibliography. A large and thorough subject index enhances the book's value as a quick reference.

A book of this kind is likely to contain errors and omissions. We have tried our best to obtain permission for published facts and figures from the original authors. If any omissions have been made, these will definitely be rectified in subsequent editions. We particularly welcome suggestions as to typing errors and additions or the deletion of facts to this book from the readers. They can mail their comments to either of the authors at the following e-mail.

J. Mishra : mishrajibitesh@gmail.com
S.N.Mishra: sarojmishra1@rediffmail.com

We take this opportunity to gratefully acknowledge the advice of Prof. A.K.Bisoi, Professor and Head, Department of Computer Sc, Utkal University, Vani Vihar, Bhubaneswar, Orissa who has aided us during research on these new concepts on fractals.

Finally, we would like to thank our respective families for their love and constant inspiration all through the process of research and writing of this book. Without their untiring zeal and enthusiasm, this book could not have been completed.

J.M
S.N.M

Dr. Jibitesh Mishra
Assistant Professor

Department of Computer Sc.& Applications,

College of Engineering & Technology, Bhubaneswar, Orissa

Dr. Sarojananda Mishra
Assistant Professor

Department of Computer Sc. & Applications,

Indira Gandhi Institute of Technology, Sarang, Dhenkanal, Orissa

Contents

Chapter 1

Introduction to Fractals

"Fractal" is a term coined by Benoit Mandelbrot in 1924 to describe an object, which has partial dimension. For example, a point is a zero-dimensional object, a line is a one-dimensional object, and a plane is a two-dimensional object. But what about a line with a kink in it? Or a line that has an infinite number of kinks in it? These are mathematical constructs which don't fit into normal (Euclidean) geometry very well, and for a long time mathematicians considered things like these "monsters" to be avoided - lines of thought that defied rational explanation in known terms. Within the past few decades, "fractal maths" has exploded, and now there are "known terms" for describing objects which were indescribable or inexplicable before. There are an infinite variety of fractals and types. Gaston Julia (1893-1978) was a French mathematician whose work (published in 1918) inspired Mandelbrot. In 1977, the second time Mandelbrot looked at Julia's work [18,32,60,62]), he used computers to explore it and discovered (quite by accident) the most famous fractal of all, which now bears his name: the Mandelbrot set.

Fractals are famous for their beauty and fractal techniques are employed as they require less space for storage. Mandelbrot [21,22,24] introduced the term fractal in 1975 to describe the shape and appearance of objects. The fractals contain their own scale down, rotate and skew replicas embedded in them. Fractal objects, as defined by Mandelbrot have certain special properties such as self-similarity (containing replicas/copies of themselves), underivability at every point, and a Hausdorff-Besicovitch (HB) dimension greater than their geometric dimension.

Most of the natural objects, at the macro-level as well as at the micro-level would better be called fractal objects rather than a combination of Euclidean primitives. That is why fractal analysis is preferred in most of the applications where natural objects are dealt with.

Many natural objects and man-made processes exhibit intricate detail and scale invariance. Given some underlying geometry, one can study these objects and processes under a branch of mathematics known as fractal geometry. Benoit Mandelbrot christened this field in 1970s. Mandelbrot's The fractal Geometry of Nature, was the first "text book" in the field.

1.1 Overview of fractals

Fractal geometry [63,64] has become a common tool to describe objects or phenomena in which a scale invariance of some sort exists. The variety of scientific applications is enormous; as such structures exist from physics to astrophysics, from biology to chemistry, and even in market fluctuations analysis. However, till recently, fractal concepts were used more to understand than to build. The period is now appropriate to stress the importance of the practical applications of fractal objects and fractal concepts in many fields of direct importance to industry, communications, environment, and physiology.

Fractals have been generated or represented by different means, like recursive mathematical families of equations, recursive transformations (generators) applied to an initial shape (the initiator) and fractional Brownian movements (fBM) [25,85].
This book chiefly discusses the initiator and generator principle. It also provides a brief introduction to fractal theory and the generation and transportation of self-similar figures through mathematical formulae.

1.2 Fractals vs. Computer Graphics

One of the earliest applications of fractals was computer graphics. It would take a lot of space to store the exact data needed to construct the cratered surface of the moon. This would be memory, if all you wanted were a realistic lunar landscape for a science fiction film. The answer to this is fractal forgeries; these mimic the desired forms without worrying about the precise details, they also require very little storage space because of easy compression.

Fractal landscapes were published in numbers in Mandelbrot's Fractal geometry of Nature. Fractals have been used in films such as Star Trek II: The Wrath of Kahn, for the landscapes of the Genesis planet, and in Return of the Jedi to create the geography of the moons of Endor and the outlines of the Death Star.

Peter Oppenheimer has used fractals to produce abstract art, as well as stylistic, yet lifelike trees and plants. Barnsley also discovered that simple fractal rules could generate intricate 'forgeries' of trees and ferns, as these objects are self-similar. The fern is built from four transformed copies of itself. The fourth is the stem which is a fern squashed flat to form a line. All that is needed to construct the fern are the rules.

These days computer-generated fractal patterns are everywhere, from squiggly designs on computer art posters to illustrations in the most serious of physics journals. Interest continues to grow among scientists and, rather surprisingly, artists and designers. This book provides visual demonstrations of complicated and beautiful structures that can be constructed in systems, based on simple rules. It also presents papers on seemingly paradoxical combinations of randomness and structure in systems of mathematical, physical, biological, electrical, chemical, and artistic interest. Topics include: iteration, cellular automata, bifurcation maps, fractals, dynamical systems, patterns of nature created through simple rules, and aesthetic graphics drawn from the universe of mathematics and art. Chaos and Fractals is divided into six parts: Geometry and Nature; Attractors; Cellular Automata, Gaskets, and Koch Curves; Mandelbrot, Julia and Other Complex Maps; Iterated Function Systems; and Computer Art. Additional information on the latest practical applications and uses of fractals in commercial products such as the antennas and reaction vessels is presented. In short, fractals are increasingly finding application in practical products where computer graphics and simulations are integral to the design process.

1.2.1 Chaotic Systems

The classic Mandelbrot below is the image that had greatly popularised chaotic and fractal systems. The Mandelbrot set is created by a general technique where a function of the form $z_{n+1} = f(z_n)$ is used to create a series of a complex variable. In the case of the Mandelbrot the function is $f(z_n) = z_n^2 + z_0$. This series is generated for every initial point z_0 on some partition of the complex plane. To draw an image on a computer screen the point under consideration is coloured depending on the behavior of the series, which will act in one of the following ways:
decay to 0
tend to infinity
oscillate among a number of states
exhibit no discernible pattern
    In the figure (Fig.1.1) below, situation (a) occurs in the interior portion, (b) in the exterior, (c) and (d) near the boundary. The boundary of the set exhibits infinite detail and variation (the boundary will never appear smooth irrespective of the zoom factor), as well as self similarity.

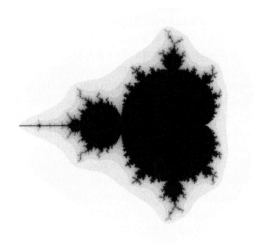

Figure 1.1: Mandelbrot Set reproduced with permission from B.B.Mandelbrot from his book *Fractal Geometry of Nature* published by W.H.Freeman

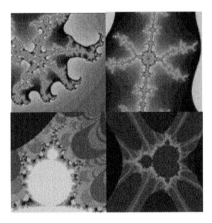

Figure.1.2: Biomorphs reproduced with permission from B.B.Mandelbrot from his book *Fractal Geometry of Nature* published by W.H.Freeman

C.A.Pickover calls an example using the same technique but a different function "biomorphs" (Fig 1.2). It uses the function f $(z_n) = \sin(z_n) + e^z + c$ and gives rise to many biological looking creatures depending on the value of the constant "c".

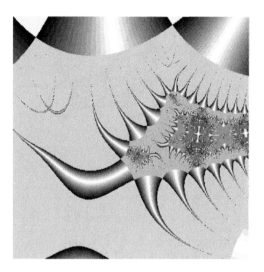

Figure.1.3: Julia Set reproduced with permission from B.B.Mandelbrot from his book
Fractal Geometry of Nature published by W.H.Freeman

1.2.2   Strange Attractor

A second technique, often called "hopalong", is normally used to represent the strange attractor of a chaotic system, for example, the well-known Julia set (Fig. 1.3). In this case each coordinate generated by the series is drawn as a small point, i.e., we hop-along from one point to the next. For an image on a plane the series is either an equation of a complex variable or there are two interrelated equations, one for the x and one for the y coordinate. As an example consider the following function:

$$Y_{n+1} = a - x_n$$

This series of x,y coordinates is specified by an initial point x_o, y_o and three constants a,b, and c. The following is an example where a=0.4, b=1, and c=0.

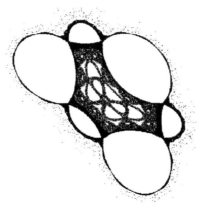

Figure 1.4: Strange attractors

Interestingly, for strange attractors the initial point does not matter (except for a few special cases), i.e., all initial coordinates x_0, y_0 result in the same image. In other words, the image shows the x, y pairs that can be generated by the series; Any initial point will generate the same set of points although they will be generated in a different order. Another example attributed to Peter de Jong uses the two equations

$x_{n+1} = \sin(a\, y_n) - \cos(b\, x_n)$
$y_{n+1} = \sin(c\, x_n) - \cos(d\, y_n)$

This gives swirling tendrils that appear three dimensional; An example is shown below where a = -2.24, b = -0.65, c = 0.43, d = -2.43.

Figure 1.5: Swirling tendrils

1.2.3 Newton Raphson

This technique is based on the Newton Raphson method of finding the solution (roots) to a polynomial equation of the form

$$f(z) = a_0 + a_1 z + a_2 z^2 + \ldots + a_m z^m = 0$$

The method generates a series where the n+1'th approximation to the solution is given by

$$z_{n+1} = z_n - f(z_n) / f'(z_n)$$

Where $f'(z_n)$ is the slope (first derivative) of f (z) evaluated at z_n. To create a 2D image using this technique each point in a partition of the plane is used as initial guess, z_o, to the solution. The point is coloured depending on which solution is found and/or how long it took to arrive at the solution. A simple example is an application of the above to find the three roots of the polynomial z*z*z - 1 = 0. The following shows the appearance of a small portion of the positive real and imaginary quadrant of the complex plane. A trademark of chaotic systems is that very similar initial conditions can give rise to very different behaviour. In the image shown there are points very close together. One of them converges to the solution very fast while the other converges very slowly.

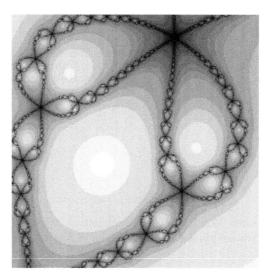

Figure 1.6: Conversion of Newton Raphson method reproduced with permission from B.B.Mandelbrot from his book *Fractal Geometry of Nature* published by W.H.Freeman

1.3 Fractal Geometry

Almost all geometric forms used for building manmade objects belong to Euclidean geometry; they are comprised of lines, planes, rectangular volumes, arcs, cylinders, spheres, etc. These elements can be classified as belonging to an integer dimension, 1, 2, or 3. This concept of dimension can be described both intuitively and mathematically. Intuitively we say that a line is one-dimensional because it only takes 1 number to uniquely define any point on it. That one number could be the distance from the start of the line. This applies equally well to the circumference of a circle, a curve, or the boundary of any object.

Any point "a" on a one dimensional curve can be represented by one number, the distance d from the start point.

Figure 1.7: Concept of Euclidean geometry

A plane is two-dimensional since in order to uniquely define any point on its surface we require two numbers. There are many ways to arrange the definition of these two numbers but we normally create an orthogonal coordinate system. Other examples of two-dimensional objects are the surface of a sphere or an arbitrary twisted plane.

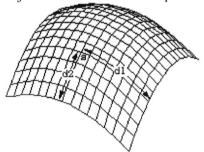

Any point "a" on a two dimensional surface can be uniquely represented by two numbers. One of the many possible methods is to grid the surface and measure two distances along the grid lines.

Figure 1.8: Example of twisted plane

The volume of some solid object is 3 dimensional on the same basis as above, it takes three numbers to uniquely define any point within the object.

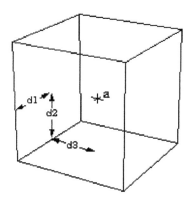

Any point "a" in three dimensions can be uniquely represented by three numbers. Typically these three numbers are the coordinates of the point using an orthogonal coordinate system.

Figure 1.9: Orthogonal coordinate system

A more mathematical description of dimension is based on how the "size" of an object behaves as the linear dimension increases. In one dimension consider a line segment. If the linear dimension of the line segment is doubled then obviously the length (characteristic size) of the line has doubled. In two dimensions, if the linear dimensions of a rectangle for example is doubled then the characteristic size, the area, increases by a factor of 4. In three dimensions if the linear dimension of a box are doubled then the volume increases by a factor of 8. This relationship between dimension D, linear scaling L and the resulting increase in size S can be generalized and written as
$$S = L^D$$
This is just telling us mathematically what we know from everyday experience. For example, if we scale a two-dimensional object, then the area increases by the square of the scaling. If we scale a three dimensional object the volume increases by the cube of the scale factor. Rearranging the above gives an expression for dimension depending on how the size changes as a function of linear scaling, namely
$$D = \log(S)/\log(L)$$
In the examples above the value of D is an integer, either 1, 2, or 3, depending on the dimension of the geometry. This relationship holds for all Euclidean shapes. There are however many shapes which do not conform to the integer based idea of dimension given above in both the intuitive and mathematical descriptions. That is, there are objects, which appear to be curves for example, but a point on the curve cannot be uniquely described with just one number. If the earlier scaling formulation for dimension is applied the formula, it does not yield an integer. There are shapes that lie in a plane but if they are linearly scaled by a factor L, the area does not increase by L squared but by some non-integer amount. These geometries are called fractals. One of the simpler

fractal shapes is the von Koch snowflake. The method of creating this shape is to repeatedly replace each line segment with the following 4 line segments.

Fig 1.10: Initiator of Koch curve

The process starts with a single line segment and continues for ever. The first few iterations of this procedure are shown below.

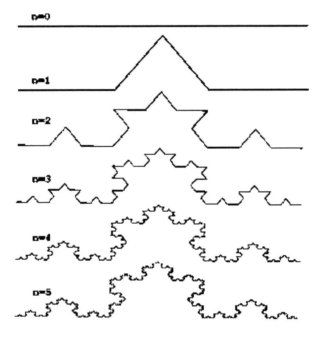

Figure. 1.11: Successive iterative generations of Koch curves

This demonstrates how a very simple generation rule for this shape can create some unusual (fractal) properties. Unlike Euclidean shapes this object has detail at all levels. If one magnifies a Euclidean shape such as the circumference of a circle it becomes a different shape, namely a straight line. If we magnify this fractal, an increasing amount of detail is uncovered, which is self-similar, in fact, exactly self-similar. In other words, any magnified portion is identical to any other magnified portion in such figures.

Also note that the "curve" on the right is not a fractal but only an approximation of one. This is no different from when one draws a circle; it is only an approximation of a perfect circle. At each iteration the length of the curve increases by a factor of 4/3. Thus the limiting curve is of infinite length and indeed the length between any two points of the curve is infinite. This curve manages to compress an infinite length into a finite area of the plane without intersecting itself. Considering the intuitive notion of 1 dimensional shape, although this object appears to be a curve with one starting point and one end point, it is not possible to uniquely specify any position along the curve with one number as we expect to be able to do with Euclidean curves, which are 1 dimensional. Although the method of creating this curve is straightforward, there is no algebraic formula that can describe the points on the curve. Some of the major differences between fractal and Euclidean geometry are outlined in the following table(Table 1.1).

Table 1.1: Comparison of fractal and Euclidean geometry

Fractal	Euclidean
Modern Invention	Traditional
No specific size or scale	Based on characteristics size or scale
Appropriate for Geometry in Nature	Suits description of man made objects
Described by an algorithm	Described by a usually simple formula

First, the recognition of fractal is very modern, they have only formally been studied in the last 10 years compared to Euclidean geometry which goes back over 2000 years. Second, whereas Euclidean shapes normally have a few characteristic sizes or length scales (e.g., the radius of a circle or the length of a side of a cube) fractals have no characteristic sizes. Fractal shapes are self-similar [68] and independent of size or scaling. Third, Euclidean geometry provides a good description of man-made objects whereas fractals are required for a representation of naturally occurring geometries. It is likely that this limitation of our traditional language of shape is responsible for the striking difference between mass produced objects and natural shapes. Finally, Euclidean geometries are defined by algebraic formulae, for example
$x^2+y^2+z^2=r^2$
defines a sphere. Fractals are normally the result of a iterative or recursive construction of algorithm.

1.4 Categories of Fractals

It is convenient to resort to conventional classification for representation of all the variety of fractals.

1.4.1   Geometrical fractals

The fractals of this class are visual [76,78,79,80]. In two-dimensional case they are made of a broken line (or of a surface in three-dimensional case) known as the generator. A broken line generator at the corresponding scale for a step of algorithm replaces each of the segments, which form the broken line. As a result of infinite repeatition of steps, a geometrical fractal is produced.

The process of construction begins from a segment of single length (Fig.1.11). This is the zero generation of the Koch curve. Then each section (one segment in zero generation) is replaced by formative element defined on the figure as **n=1**. As a result of the substitution we get the next generation of the Koch curve. There are four rectilinear sections **1/3** length when the first generation is produced. Thus, to produce the next generation all of the sections of previous generation are replaced by diminished formative element. The curve of **n-th** generation is called prefractal when **n** is finite quantity. When **n** is infinite, the curve is considered a fractal object.

To make another fractal object we have to change the rules of construction process. Let the formative element be two equal segments combined at the right angle. In zero generation a single segment is replaced by the formative element such a way that the angle is from above. So there is a removal of the middle of the section. To get the next generation we replace the first section from the left by the formative element in such a way that the middle of the section shifts to the left from the direction of movement. To replace the next sections we have to alternate the directions of middle shifts. Fig 1.12 shows several generations and 11th generation of the curve constructed in the same manner. When **n** is infinite, the fractal curve is considered Harter-Heituey's "dragon".

Figure 1.12: The Harter-Heituey's "dragon" reproduced with permission from
B.B.Mandelbrot from his book *Fractal Geometry of Nature* published by W.H.Freeman.

1.4.2 Algebraic fractals

Algebraic fractals is the biggest class of fractals. Using nonlinear processes in n-dimentional spaces creates them. Two-dimensional processes are mostly scrutinized. Interpreting nonlinear iterate process as a discrete dynamical system we can use the terminology of dynamical system theory.

It is known that nonlinear dynamical systems possess several states. The state in which the dynamic system is after several iterations depends on its initial state. Hence, in each of the states so-called attractor possesses of some area of initial states from which system undoubtedly hits into considered one of finite states. Thus, phase space of system is divided into areas of attraction. If two-dimensional space is phase painting areas of attraction with different colours we can get colour phase portrait of the system (iteration process). Changing the algorithm of choosing the colours, we can get complicated fractal images with fantastical multi-colour traceries. Mathematicians had not expected to find a chance to create such nontrivial structures using primitive algorithms.

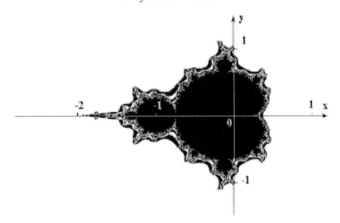

Figure 1.13: The Mandelbrot set reproduced with permission from B.B.Mandelbrot from his book *Fractal Geometry of Nature* published by W.H.Freeman

Let us take up the Mandelbrot set as an example (Look at Fig.1.13 and Fig.1.14). The construction algorithm is rather simple and based upon plain iterate expression:

$$\mathbf{Z}\,[i+1] = \mathbf{Z}\,[i] * \mathbf{Z}\,[i] + \mathbf{C},$$

where $\mathbf{Z}i$ and $\mathbf{C}$ are complex variables. Iterations implement for each start point $\mathbf{C}$ of rectangular or square area, which is subset of complex plane. Iterate process continues so long as $\mathbf{Z}\,[i]$ is not out of circle with radius 2 center of which is in point (0,0) (it means that attractor of dynamical system is in infinity) or so long as after rather large quantity of iterations (for instance 200-500) $\mathbf{Z}\,[i]$ does not coincide with some point on the circle. According to quantity of iterations during of those $\mathbf{Z}\,[i]$ was inside the circle one can set the colour of point $\mathbf{C}$. (if $\mathbf{Z}\,[i]$ has been inside the circle for a large number of iterations, the iterate process stops and the point has to be coloured in black).

The aforementioned algorithm gives the approximation to the Mandelbrot set. The Mandelbrot set has points, which do not go off to infinity during infinite quantity of iterations (which are black). Points, which are on the bound of the Mandelbrot set, go off to infinity during finite quantity of iterations.

Figure 1.14: Section of enlarged Mandelbrot set bound reproduced with permission from B.B.Mandelbrot from his book *Fractal Geometry of Nature* published by W.H.Freeman

1.4.3   Stochastic fractals

The stochastic fractals are generated if the iterate process has accidental parameters. Objects similar to those found in nature can be created by using these methods. Two-dimensional stochastic fractals are used for designing the surface of the sea or for relief modeling.

There are other classifications of fractals. Fractals can be divided into determine (algebraic and geometric) and nondetermine (stochastic). The deterministic fractals can be redivided into linear and nonlinear fractals.

1.5 Fractals and Non-fractal Objects

Non-Fractal

As a **non-fractal** object is magnified, no new features are revealed.

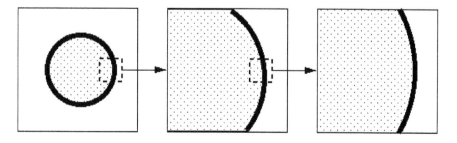

Figure 1.15: Non-fractal objects

L-System Fractals

Fractal

As a **fractal** object is magnified, ever finer new features are revealed. The shapes of the smaller features are kind-of-like the shapes of the larger features (Fig 1.16).

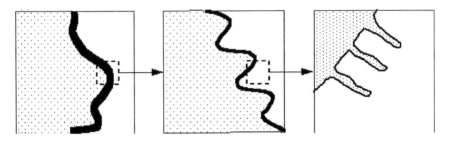

Figure 1.16: Fractal objects

1.5.1   The sizes of the features of the fractal and non-fractal objects

Non-Fractal

The size of the smallest feature of a **non-fractal** object is called its *characteristic scale*. When we measure the length, area, and volume at resolution that is finer than the characteristic scale, then all of the features of the object are included. Thus, the measurements at this resolution determine the correct values of the length, area, and volume.

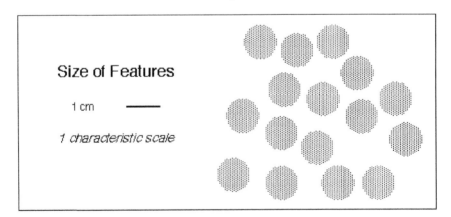

Figure 1.17: Sizes of features of non-Fractals

Fractal

A **fractal** object has features over broad ranges of sizes. There is no single *characteristic scale*. As we measure the length, area, and volume at ever finer resolution, we include ever more of its finer features. Thus, the length, area, and volume depend on the resolution used to make the measurement.

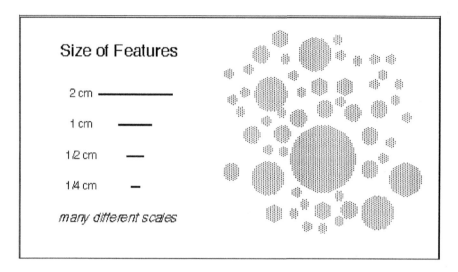

Figure 1.18: Features of fractals

1.5.2 The four major properties of fractals

Self-Similarity
A coastline looks wiggly. You would think that as you enlarge a piece of the coastline the wiggles would be resolved and the coastline would look smooth. But it does not. No matter how much you enlarge the coastline it still looks just as wiggly. The coastline is similar to itself at different magnifications. This is called **self-similarity**.

Scaling
Because of self-similarity, features at one spatial resolution are related to features at other spatial resolutions. The smaller features are smaller copies of the larger features. The length measured at finer resolution will be longer because it includes these finer features. How the measured properties depend on the resolution used to make the measurement is called the **scaling relationship**.

Dimension
The dimension gives a quantitative measure of self-similarity and scaling.
It tells us how many new pieces of an object are revealed as it is viewed
at higher magnification.

Statistical Properties
Most likely, the statistics that you were taught in school was limited to
the statistics of non-fractal objects. Fractals have different statistical
properties.

1.6 Defining a fractal

Fractals are a relatively new field of mathematics. Though basic theories
were put forth in the early 20th century, no real fractal was created and
studied until the 1970's when Benoit B. Mandelbrot put forth the first
iterative formula for generating a fractal. This man alone can be credited
with defining the field of study, which he himself named fractals.
Today, fractal mathematics is a burgeoning field of study. Interest in the
topic has spread through the mathematical community and into popular
culture. Fractals are applied in today's art, media, communications
industry, and economical studies.

1.6.1   Definitions of related terms

Before we learn how to generate a fractal, we must know some
definitions. The following terms are key in determining if you indeed
have a fractal.

Bounded
A set in R^2 is bounded if a suitably large circle can enclose it.

Closed
If a set contains all of its boundary points, it is said to be closed.

Congruent
Two sets in R^2 are congruent if they can be made to coincide exactly by
translating and rotating them appropriately within R^2.

Self-similar
A set $S \in R^2$ is self similar if S is closed and bounded and can be
expressed in the form
    $S = S1 \cup S2 \cup S3 \cup ... \cup Sk$

Where S1, S2, S3,..., Sk are nonoverlapping sets, each of which is congruent to S scaled by the same factor s (0<s<1).

Topological dimension, d_T
While the general definition requires an understanding of topology, for this book, it will suffice to outline some conclusions from this definition:

A point in R^2 has topological dimension zero
A curve in R^2 has topological dimension one
A region in R^2 has topological dimension two

Hausdorff dimension, d_H
The Hausdorff dimension, $d_H(S)$, of a self-similar set S is
$$d_H(S) = \ln(k)/\ln(1/s)$$

Where k is the number of nonoverlapping sets composing S and s is the scale factor.
Note that $d_H(S)$ does not need to be an integer and that
$d_T(S) \leq d_H(S)$

1.6.2   Definition of fractal

A fractal is a subset of a Euclidean space whose Hausdorff dimension and topological dimension are not equal.

Madelbrot's definition
A rough or fragmented geometric shape that can be subdivided in parts, each of which is (at least approximately) a smaller sized copy of the whole.

Mathematical definition
A set of points whose fractal dimension exceeds its topological dimension.

1.7 Applications for Fractals

The most amazing thing about fractals is the variety of their applications. Besides theory, they were used to compress data in the Encarta Encyclopedia and to create realistic landscapes in several movies like Star Trek. The places where you can find fractals include almost every part of the universe, from bacteria cultures to galaxies.

Even though this field of mathematics is still in its infancy, it has already seen real life applications in a vast array of fields. The following are just a few of its many applications. Motorola has started using fractal antennae in many of its cellular phones, and reports that they're 25% more efficient than the traditional piece of wire. They're also cheaper to manufacture, can operate on multiple bands, and can be put into the body of the phone.

Fractals are used in art, movie special effects, simulation of natural scenes [29,35,44,49,58,72,81,131] and video games. Art takes advantage of the aesthetic appeal of fractal images, while movies and video games also benefit by the fact that fractal images are easy to compress and store without taking a lot of memory. Instead of having to store each point of an image, only a seed and the similitude equations need be stored. Below is one of the most famous examples of a fractal used in a movie Return of the Jedi (death star).

After an introduction by Benoit Mandelbrot on "The saga of *fractals*, from practical questions, to geometry, and back to *applications*" many themes of practical importance have gathered considerable attention in various areas, given below in categories:

Fractals in industry and man-made *fractals*:

Fractal antennae
Fractal sound barriers
Use of fractal polymeric surfaces
Fractal reactor design
Fractal studies of heterogeneous catalysis
Petroleum research

Natural fractal objects:

Fractal bronchial trees in mammals
Growth of fractal trees in nature
Optimal fractal distribution
Absolute limitations of tree distributive structures
River Networks

Applications of fractal concepts to the study of complex systems:

Image analysis and compression
Multifractal signal analysis
Scaling topology of the Internet and the www
Fractal aviation communication network

1.8 Summary

The real application of much of the above has arisen from attempts to model natural phenomena in the world we live in. Many of the mathematical techniques have found a firm place in the computer graphics industry as a means of creating both stunning graphical images as well as very natural looking structures. As the techniques become more standardized and more application areas are found they are likely to be incorporated as one of the standard tools in CAD, painting and image processing software packages. Fractals are one of the most important discoveries in mathematics of the last century. Benoit Mandelbrot managed to pull together all the scattered concepts that had been thought up over the years from previous mathematicians such as Julia and Koch, and build a concrete foundation for this new branch of mathematics.

The unique properties of fractals have lead to widespread interest in trying to harness the power of these complex geometric objects. The bewildering simplicity of the processes used to create these images, begs the question whether their true potential can ever be fully realised. It is as if they are almost organic in structure, somehow knowing how to define themselves.

Chapter 2

Fractals and L-System

L-systems are a mathematical formalism proposed by the biologist Aristid Lindenmayer in 1968 as a foundation for an axiomatic theory of biological development. Two principal applications of L-systems include the generation of fractals and the realistic modeling of plants.

Central to L-systems is the notion of rewriting, where the basic idea is to define complex objects by successively replacing parts of a simple object using a set of rewriting rules or productions. The rewriting can be carried out recursively.

The most extensively studied and the best understood rewriting systems operate on character strings. Chomsky's work on formal grammars spawned a wide interest in rewriting systems. Subsequently, a period of fascination with syntax, grammars and their application in computer science began, giving birth to the field of formal languages.

Aristid Lindenmayer's work introduced a new type of string rewriting mechanism, subsequently termed L-systems. The essential difference between Chomsky grammars and L-systems lies in the method of applying productions. In Chomsky grammars [124] productions are applied sequentially, whereas in L-systems they are applied in parallel, replacing simultaneously all letters in a given word. This reflets the similarity of L-systems with biological processes. Productions are intended to capture cell divisions in multicellular organisms, where much division may occur at the same time.

2.1 Reviews on L-systems

In 1968, Aristid Lindenmayer introduced L-systems, which provided a mathematical formalism for parallel grammars well adapted to the modeling of growth phenomena [13,112]. In 1984, Alvy Ray Smith, a computer graphics researcher showed how L-systems could be used to synthesize realistic images. He also pointed out the relationship between the concept of Fractals and L-systems [13,14,15]. L-systems used to generate plants with or without inflorescence, cell growth and geometric patterns such as Indian kolams or mathematical 'monsters' such as the Von Koch or Hilbert curves. Many geometric patterns and tilings can be generated using L-systems. However, the problem of describing patterns and tilings using L-systems is largely unexplored.

In 1988, Friedell and Schulmann developed a prototype for the automatic generation of architectural scenes which would allow the infinite range of forms generated using L-systems formalism to be explored.

One set of rules (in the simplest example, only one rule) describes the pattern to be repeated.

Another rule describes the expansion of the pattern through translations. And a third rule describes the rotational symmetries. Using the basic structure, all symmetry groups can be described easily using the same grammar skeleton.

The rule-based language of "stochastic sensitive growth grammars" as an extension of parametric L-systems [12] was developed to describe algorithmically the change of the morphology of forest trees in time, taking endogenous and exogenous factors into account, and to create systematically three-dimensional simulations of tree crowns. With different species, mainly of trees, spruce, morphological measurements were carried out to get a basis for the design and parameterization of such rule systems.

The software GROGRA (Growth Grammar Interpreter) creates a time series of three-dimensional crown structures from the rules; the basic elements of these structures (annual shoots) can additionally bean non-geometrical attributes. Furthermore, GROGRA contains several analysis tools and data interfaces.

The generated architectures serve as an "ecomorphological basis model" for different process-oriented simulation models. There is already a realized a model of tree-internal water flow (HYDRA) based on the artificial tree structures.

Lin implemented the animation of L-systems based on three-dimensional plant growing in Java. His animation used a number of iterations to animate plant development. His animation was not smooth.

Prusinkiewicz, James, and Mech extended Lindenmayer systems [107] in a manner suitable for simulating the interaction between a developing plant and its environment. The formalism was illustrated by modeling the response of trees to pruning, which yields synthetic images of sculptured plants found in topiary gardens.

Hammel and Prusinkeiwicz extended the notation of L-systems with turtle interpretation [103,108,109,110] to facilitate the construction of such objects. The extension was based on the interpretation of the entire derivation graph generated by L-systems, as opposed to the interpretation of individual words. They illustrated the proposed method by applying it to visualize the development of compound leaves, a sea shell with a pigmentation pattern, and a filamentous bacteria.

Prusinkiewicz extended it further to language-restricted iterated function systems (LRIFS's) [103]. They generalized the original definition of IFS's by providing a means for restricting the sequences of applicable transformations. The resulting attractors include sets that cannot be generated using ordinary IFS's. Their research was expressed using the terminology of formal languages and finite automata.

Prusinkiewicz, Hammel, and Mjolsness introduced a combined discrete / continuous model of plant development that integrates L-system-style productions and differential equations [106]. The model was suitable for animating simulated developmental processes in a manner resembling time-lapse photography. The proposed techniques were illustrated using several developmental models, including the flowering plants.

Prusinkiewicz and Kari expressed the development of modular branching structures [105] that satisfy three assumptions: (a) sub apical branching, meaning that new branches can be created only near the apices of the existing branches, (b) finite number of module types and states, and (c) absence of the interactions between coexisting components of the growing structure. These assumptions were captured in the notion of subapical bracketed deterministic L-systems without interactions (sBOL-systems). They presented the biological rationale for sBOL-systems and proved that it is decidable whether a given BOL-system was subapical or not.

Hemmel, Prusinkiewicz, Remphrey, and Davidson presented a methodology for creating models that capture the development of a plant using the formalism of L-systems and incorporating biological data using *Fraxinus penmylvanical* shoots based on L-systems [103].

Hammel, Prusinkiewicz, and Wyvill proposed a method for modeling compound leaves in plants [106]. A branching skeleton generated using L-systems captures the layout of leaf lobes. The leaf margin is then traced around the skeleton. Their work focused on the specification and tracing of the margin, and included references to the techniques described in the literature for performing the other tasks. The margin is defined as an implicit contour.

Lintermann and Deussen presented a modeling method and graphical user interface for the creation of natural branching structures such as plants [51,95]. Structural and geometric information is encapsulated in objects that are combined to form a description of the model. The model was represented graphically as a structure graph and could be edited interactively. Global and partial constraint techniques were integrated on the basis of tropism, naturally occurring deformations and pruning operations to allow the modeling of specific shapes.

Deussen developed a system built around a pipeline of tools [95]. The terrain was designed using an interactive graphical editor. Plant distribution was determined by hand, by ecosystem, or by a combination of both techniques.

These two topics of literature review have stimulated the ideas of researchers for improving the previous works, which will be discussed later.

2.2 Parallel grammars: A phenomenon

From an abstract level, parallel grammars (and automata) try to model phenomena encountered in parallel processing.

Since in grammatical mechanisms for generating formal languages the basic and simplest processing step is that of a replacement of one symbol (or a string of contiguous symbols in the non-context-free case, but this can be seen as a special case of the context-free case if we consider parallel derivations restricted to adjacent symbols) by a string of symbols. The most straightforward way to incorporate or model parallelism here is certainly to consider the possibility of doing a number of derivation sub-steps in parallel—hence denoting the notion of a *parallel derivation step*. One obvious possibility here is to allow (and enforce) the parallel execution of a context-free derivation step in all possible places. This basically leads to the theory of Lindenmayer systems [12,13,42,52]. For physical reasons (and often also biologically motivated), it is unreasonable to assume that basic processing units (cells) located in the midst of a larger computing facility (body) may grow arbitrary (by "splitting" the cell due to some rule of the form $A \rightarrow BC$). This idea led, on the one hand, to the theory of cellular automata (when we, for the moment, forget about the differentiating aspects of grammars and automata) and on the other hand to considering Lindenmayer systems with apical growth [93,105,127,130]. In this context, we should also mention array grammars, which adhere to similar growth restriction, although they are not necessarily parallel in nature, but, as often the case in formal language theory, different concepts of rewriting; here the ideas of array grammars and of parallelism in rewriting can be fruitfully combined [30,111,114]. Even more interesting than the usually considered one-dimensional case are the higher-dimensional cases, where the restrictions of growth in the innermost cells of the body become even more obvious [74,75,87,114]. When comparing the growth patterns, i. e., consider the development of the length of the cellular strings as a function of time, which are typically found within (certain types of simple) Lindenmayer systems they mostly show polynomial or exponential growth.

2.3 L-Systems

L-systems are a mathematical formalism proposed by the biologist Aristid Lindenmayer in 1968 as a foundation for an axiomatic theory of biological development. More recently, L-systems have found several applications in computer graphics. Two principal areas include generation of fractals and realistic modeling of plants.

Central to L-systems, is the notion of rewriting, where the basic idea is to define complex objects by successively replacing parts of a simple object using a set of rewriting rules or productions. The rewriting can be carried out recursively.

The most extensively studied and the best understood rewriting systems operate on character strings. Chomsky's work on formal grammars spawned a wide interest in rewriting systems. Subsequently, a period of fascination with syntax, grammars and their application in computer science began, giving birth to the field of formal languages.

Aristid Lindenmayer's work introduced a new type of string rewriting mechanism, subsequently termed L-systems [13]. The essential difference between Chomsky grammars and L-systems lies in the method of applying productions. In Chomsky grammars productions are applied sequentially, whereas in L-systems they are applied in parallel, replacing simultaneously all letters in a given word. This difference reflects the biological motivation of L-systems. Productions are intended to capture cell divisions in multicellular organisms, where many divisions may occur at the same time.

2.3.1 D0L-system

The simplest class of L-systems is termed D0L-systems (D0L stands for deterministic and 0-context or context-free). To provide an intuitive understanding of the main idea behind D0L-systems, let us consider this example given by Prusinkiewicz and Lindenmayer.

Lets us consider strings built of two letters a and b (they may occur many times in a string). For each letter we specify a rewriting rule. The rule $a \rightarrow ab$ means that the letter a is to be replaced by the string ab, and the rule $b \rightarrow a$ means that the letter b is to be replaced by a. The rewriting process starts from a distinguished string called the axiom. Let us assume that it consists of a single letter b. In the first derivation step (the first step of rewriting) the axiom b is replaced by a using production $b \rightarrow a$. In the second step a is replaced by ab using the production $a \rightarrow ab$. The word ab consists of two letters, both of which are simultaneously replaced in the next derivation step. Thus, a is replaced by ab, b is replaced by a, and the string aba results. In a similar way (by the simultaneous replacement of all letters), the string *aba* yields *abaab* which in turn yields *abaababa*, then *abaababaabaab*, and so on.

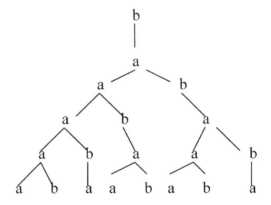

Figure 2.1: D0L System on Fibonacci number

2.3.2 Fractals and graphic interpretation of strings

Lindenmayer systems were conceived as a mathematical theory of development. Thus, geometric aspects were beyond the scope of the theory. Subsequently, several geometric interpretations of L-systems were proposed in order to turn them into a versatile tool for fractal and plant modeling.

Many fractals (or at least their finite approximations) can be thought of as sequences of primitive element -line segments. To produce fractals, strings generated by L-systems must contain the necessary information about figure geometry. Prusinkiewicz et al describe a graphic interpretation of strings, based on turtle geometry. This interpretation may be used to produce fractal images.

A state of the turtle is defined as a triplet *(x, y, a)*, where the Cartesian coordinates *(x, y)* represent the turtle's position, and the angle *a*, called the heading, is interpreted as the direction in which the turtle is facing. Given the step size *d* and the angle increment *b*, the turtle can respond to the commands represented by the following symbols:

F   Move forward a step of length d. The state
of the turtle changes to (x',y',a), where
x'= x + d cos(a) and y'= y + d sin(a). A line segment between points (x,y) and (x',y')
is drawn.

f   Move forward a step of length d without

drawing a line. The state of the turtle
changes as above.

+ Turn left by angle b. The next state of
the turtle is (x,y,a+b).

- Turn left by angle b. The next state of
the turtle is (x, y,a-b).

The turtle ignores all other symbols. The turtle preserves its current
state. Given a string v, the initial state of the turtle *(x0, y0, a0)*, and fixed
parameters d and b, the turtle interpretation of v is the figure (set of lines)
drawn by the turtle in response to the string v.

The above description gives us a simple method for mapping strings
to pictures, which may be applied to interpret strings generated by L-
systems. The next figure shows four approximations of the curve known
as "quadratic Koch island". These figures were obtained by interpreting
strings generated by the following L-system:

w: F+F+F+F
p: F → F+F-F-FF+F+F-F

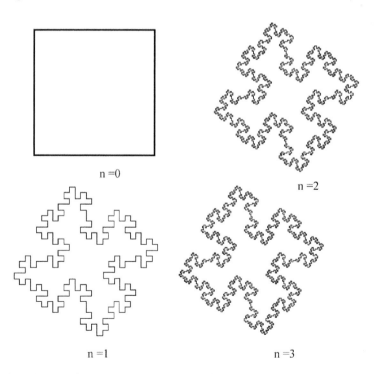

Figure 2.2: The images correspond to the strings obtained by derivations of length n = 0,
1, 2 and 3 respectively. The angle increment b is equal to 90 degrees.

2.3.3 Bracketed L-systems and models of plants architecture

Following the previous section description, the turtle interprets a character string as a sequence of line segments, connected "head to tail" to each other. Depending on the segment lengths and angles between them, the resulting figure would be more or less convoluted, but always remains just a single line.

In his work, Lindenmayer, introduced a notation for representing graph-theoretic trees using strings with brackets. The idea was to formally describe branching structures found in many plants, from algae to trees, using the framework of L-systems. Again, posterior geometric interpretations of strings with brackets were proposed for realistic modeling of plants. Thus, to represent branching structures, L-systems alphabet is extended with two new symbols, "[" and "]", to delimit a branch. The turtle interprets them as follows:

[Pop a state from the stack and make it
    the current state of the turtle.

]   Push the current state of the turtle
    onto a pushdown stack.

An example of a bracketed string and its turtle interpretation, obtained in derivations of length n = 1 - 5, are shown in the next figure. These figures were obtained by interpreting strings generated by the L-system:

w: F
p: F → F[-F]F[+F][F]

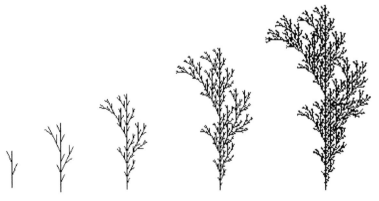

Figure 2.3: Figures from Bracketed L-System

2.3.4 L-systems and Genetic Algorithms

The fitness function employed was based on current evolutionary hypotheses concerning the factors that have had the greatest effect on plant evolution.

Figure 2.4: Figures using Genetic Algorithm

Visually appealing figures not resembling plants, were also obtained using a Genetic Algorithm with a fitness function favoring bilateral symmetric structures.

Figure 2.5: Figures using Genetic Algorithm

2.4 Basic definitions of L-Systems

Lindenmayer systems, or *L*-systems for short, are a particular type of symbolic dynamical system with the added feature of a geometrical interpretation of the evolution of the system [51]. Aristid Lindenmayer invented them in 1968 to model biological growth. The limiting geometry of even very simple systems can be extraordinary fractals.

Rather than reproduce all that, here we shall start a bit more abstractly with a symbolic dynamical system. The components of an *L*-system are as follows:

Alphabet:

The *alphabet* is a finite set V of formal symbols, usually taken to be letters a, b, c, etc., or possibly some other characters.

Axiom:

The *axiom* (also called the *initiator*) is a string w of symbols from V. The set of strings (also called *words*) from v is denoted v^*. Given $v=\{a,b,c\}$, some examples of words are *aabca, caab, b, bbc*, etc. The *length* $|w|$ of a word w is the number of symbols in the word.

Productions:

A *production* (or *rewriting rule*) is a mapping of a symbol $a \in v$ to a word $\omega \in v^*$. This will be labelled and written with notation: $p : a \rightarrow w$.

We allow as possible productions mappings of a to the *empty word*, denoted ϕ, or to a itself. If a symbol does not have an explicitly given production, we assume it is mapped to itself by default. In that case, a is a *constant* of the *L*-system.

2.4.1 Fibonacci L-system

Our first example will be the *Fibonacci L-system* with the following components

  $V = \{a,b\}$

  $\omega = a$

  $p_1 = a \rightarrow b$

  $p_2 = b \rightarrow ba$

The fascinating behavior happens when we set an *L*-system in motion, *evolving* from moment to moment. The *evolution* of an *L*-system is defined as a sequence $\{g_n\}$, $n = 0,1,2,3\dots$ where each *generation* g_n is a word in v^* that evolves from the previous generation g_{n-1} by applying all the production rules to each symbol in g_{n-1}. The first generation g_0 is the axiom ω. The first few generations of the Fibonacci system are as follows:

We may think of the individual a's and b's as life-forms, and the productions as stages in their lives. After one generation, the "immature" life-form a matures into an adult b. After that the adult b is able to produce one baby a each generation. Thus, this symbolic dynamical system models a very simple kind of population dynamics.

The origin of the name "Fibonacci" comes from the connection with a special sequence of numbers $1,1,2,3,5,8\dots$ known as the Fibonacci numbers. A second-order recurrence relation defines these

  $F_0 = F_1 = 1$

$F_{n+2} = F_{n+1} + F_n$ for $n \geq 0$

Table 2.1: First seven generations

g_0	a
g_1	b
g_2	ba
g_3	bab
g_4	$babba$
g_5	$babbabab$
g_6	$babbababbabba$
g_7	$babbababbabbababbabab$

It may be checked for the few generations listed above that the number of organisms in the n-th generation g_n is $\mid g_n \mid = F_n$. Why this should be true is a mystery until we think about it further. The reason is that this dynamical system, defined by a *local process* affecting each individual organism (the production rules), exhibits a *global process* at work at the same time, defined in terms of whole populations. After staring at the generations, we see the pattern: $g_{n+2} = g_{n+1}g_n$ by which we mean that generation *n*+2 consists of generation *n*+1 followed by generation *n*, for example:

$g_7 = $ (babbababbabba)(babbabab)$= g_6\ g_5$

2.4.2 Types of L-systems

The global process for the Fibonacci system works because this *L*-system is *context-free* meaning that the production rules take account only of an individual symbol, and not of what their neighbors are. It is possible to consider *context-sensitive* *L*-systems [66] where the production rules apply to a particular symbol only if the symbol has certain neighbors. Here is an example:

$V = \{a,b,c\}$

$\omega = bbb$

$p_1 : a(>0) \rightarrow b$

$p_2 : a(>b) \rightarrow 0$

$p_3 : b(>a) \rightarrow c$

$p_4 : b(>b) \rightarrow ba$

$p_5 : c \rightarrow 0$

This describes a population with an elaborate dynamics. If an *a* has nothing on the right (this is what the > 0 notation means), it matures to a *b*. If it has an adult *b* on the right, the adult "kills" it and it disappears. If an adult *b* has a child *a* on the right side, it grows "old" and becomes a *c*. By production p_4, if two *b*'s occur consecutively, they reproduce and produce an *a* in between. An old organism *c* dies after one generation. In all other cases the symbols stay the same. The evolution of this *L*-system is as follows:

Table 2.2: Population growth and decline

g_0	*bb*
g_1	*babab*
g_2	*ccb*
g_3	*b*

After this point the generations are unchanging. We have not stipulated yet that for each symbol in the alphabet there is exactly one production, although this has been true for our first few examples. If there is indeed exactly one production for each symbol, then the *L*-system is called *deterministic* and the sequence of generations g_n is uniquely defined as a sequence of elements of v^*. If there is more than one production for a given symbol say a $\rightarrow \omega_1$ and a $\rightarrow \omega_2$, we need a criterion for deciding when to apply which rule. One possibility is to use one of the possible productions with certain probabilities. This is called a *stochastic L-system* (the word stochastic always connotes an element of randomness [1,11,16,17,23]). In this section, we will consider only deterministic *L*-systems. A deterministic context-free *L*-system is popularly called a *D0L-system* [2,3].

The study of generation of formal subsets of v^* by means of processes similar to that underlying *L*-systems originated in formal language theory, advanced by Chomsky as a mathematical way for discussing the formation and evolution of natural languages. For this reason, any subset *S* of v^* is called a *language*. *L*-systems languages are examples of a much broader class known in theoretical computer science as *recursively enumerable* languages. Some novel aspects of *L*-systems are the parallel nature of the evolution of one word from the next and the dynamic nature in that we can think of the language as growing over time.

The Fibonacci global process could be considered a simple example of what is known as *emergent behavior* in dynamical systems theory, behavior that is apparent as the system evolves. We will see this sort of phenomenon at various times in the course. This particular global process is a *chain* or *concatenation law*, in that several past generations are chained together to form the next generation. In general, we say the *L*-system satisfies a chain law (or is *locally catenative*) if there is a sequence of integers

$0 < i_1 < i_2 < \ldots\ldots\ldots < i_k$, such that

$g_n = g_{n-i1} g_{n-i2} \ldots\ldots\ldots g_{n-ik}$

If this holds for one *n*, the same argument as that underlying the growth process for the Fibonacci system proves that it holds for all subsequent generations.

2.4.3 Thue-Morse L-system

As our next example of an *L*-system, we introduce the *Thue-Morse* system:

$V = \{a,b\}$

$\omega = a$

$p_1 = a \rightarrow ab$

$p_2 = b \rightarrow ba$

The sequence of generations starts as follows:

Table 2.3: Thue-Morse generations

G_0	a
G_1	ab
G_2	abba
G_3	abbabaab
G_4	abbabaabbababba

The number of organisms clearly doubles each generation, so that $|g_n| = 2^n$. The first question we might ask is what is the global process that emerges, if any.

To answer this question, we first observe that generation g_n does appear at the beginning of $|g_{n+1}|$. If we look at what follows, we eventually conclude that it resembles g_n, except that each *a* has been replaced by a *b* and each *b* has been replaced by an *a*. For any word *w* in v^*, we define $R(w)$ to be the mirror image word where we replace *a* by *b* and *b* by *a*. Then the global process is

$g_{n+1} = g_n R(g_n)$

Since g_n is the beginning of g_{n+1}, this *L*-system produces in the limit an infinite sequence g_∞ of *a*'s and *b*'s that begins:

g_∞ = abbabaabbababba..............

2.4.4 Paper folding and the Dragon curve

As we have stated earlier, one of the original goals of introducing *L*-systems was to model growth of various kinds. To do that, we should attach some natural meaning to the formal symbols, which are manipulated in the *L*-system. In this section, we will begin to see how the production rules can have geometric interpretations; the resulting *L*-systems can produce wonderful geometric objects: various kinds of *fractal* curves, shapes that mimic the natural world, and tilings of various kinds.

We will start with a very simple *L*-system:

$V = \{a,b\}$

$\omega = a$

$p_1 = a \to ab$

$p_2 = b \to ab$

The generations simply double each time: *a, ab, abab, abababab*, etc. To *a* and *b*, we associate the shapes:

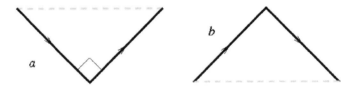

Figure 2.6: *a* and *b* triangles

The solid black lines indicate the curve that is drawn. The dotted blue line is for reference. The triangle is an isosceles right triangle. The productions in our *L*-system now represent a change in the geometry of these configurations. The solid black lines become dotted blue lines, while a solid black curve is drawn along smaller isosceles right triangles along the dotted blue lines. The two new configurations are

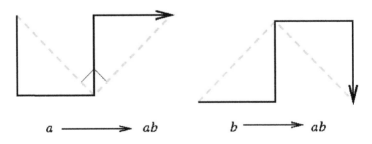

Figure 2.7: Production Rules for Dragon Curve

The first four generations of this *L*-system are displayed below, starting with the *a* triangle.

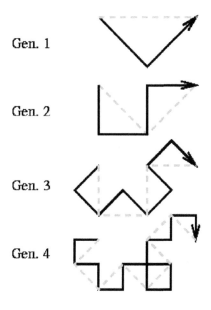

Figure 2.8: First Four Generations of Dragon Curve

As we carry out more and more generations, a wildly kinky curve known as the *Dragon* appears.

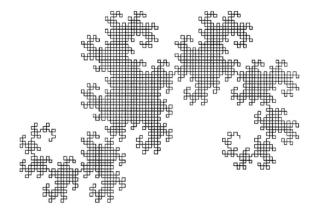

Figure 2.9: Dragon Curve 12th Generation

We may start to see some beautiful emergent behavior in this dynamical system. The curve apparently loops back to close off many small squares. These squares cluster in a sequence of "islands" each island connected to the next by a single strand. The limiting shape is the *dragon fractal*:

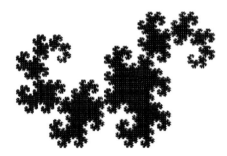

Figure 2.10: Dragon Fractal

Mandelbrot calls this the Harter-Heighway dragon. There is a natural dynamical process of "folding" at work in the creation of the dragon curves. Start with a rectangular piece of paper, which we shall view from the edge. Fold the right half over the left half, with a sharp crease down the middle. Take the folded paper and fold again the same. Continue this folding process for a few more generations. The appearance of the edge is shown on the left side of the figure below. After a number of folds, unfold the paper, and spread each fold to an angle of exactly $90°$. The resulting edge curve is our dragon. The right half of the figure shows the results for the first few generations.

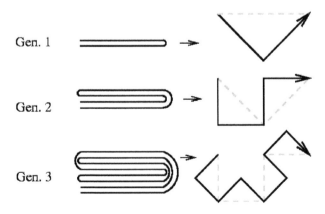

Gen. 1

Gen. 2

Gen. 3

Figure 2.11: Paper folding

There is another natural encoding of the dragons that we can see in the pictures. Following the curve from beginning to end, each turn is either to the left or to the right. Thus, each generation of the dragon corresponds to a sequences of L's (lefts) and R's (rights). In the next picture, we show generation 4 with all the turns labelled as L or R.

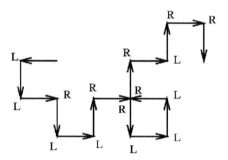

Figure 2.12: Generation 4 labeled with left and right turns

The corresponding sequences up to generation 4 are listed below.

Table 2.4: Population growth and decline

1 fold	L
2 fold	LLR
3 fold	$LLRLLRR$
4 fold	$LLRLLRRLLLRRLRR$

There is a vague similarity to the Thue-Morse sequence. Indeed, there is a relationship, but it is fairly elaborate to work out. The relation with the paper folding process suggests many interesting variations on the dragon curve. For instance, instead of always folding the right half over the left, we may occasionally fold the left half over the right. Depending on the sequence of paperfolding instructions we get different dragon curves. Strictly alternating left and right folds produces a sequence of L's and R's that arose in a question of analysis (about trigonometric sums) first considered by Rudin and Shapiro.

2.5 Turtle graphics and *L*-systems

Seymour Papert invented "Turtle graphics" [126] as a system for translating a sequence of symbols into the motions of an automaton (the "turtle") on a graphics display; the original idea was to provide a programmable object that children could learn to think geometrically with). This turned out to be an ideal system for giving a geometrical interpretation to the dynamics of *L*-systems. The basic system is as follows. We fix a "step length" *d* to be the distance covered by the turtle in one step. We also set δ to be a given angle, usually 360° / n for some integer *n*.

F:
Draw forward one step in current heading
+:
Turn heading by δ counterclockwise
-:
Turn heading by δ clockwise

As a first example of the use of these symbols, we introduce the Koch curve *L*-system:

$V = \{F,+,-\}$
$\omega = F$
$p_1 = F \rightarrow F+F\text{—}F+F$

Without specific productions mentioned for some symbols (in this case + and -), we assume they are *constants* in the sense that the implied production rule is a → a. Generation 0 is just a straight line; we will assume the heading at the start is along the positive horizontal axis. We take the angle δ to be 360°/n. Geometrically, this production means replace a straight line segment *F* by the following arrangement of four line segments.

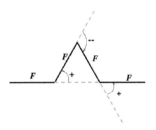

Figure 2.13: The production for the Koch curve

In this figure, the actions taken in traversing the path are marked by the corresponding symbol in *F+F-F+F*. If we scale by a factor of three so that the starting and ending points of the curve remain the same, we may describe this production by saying that the middle third of the original line segment was replaced by the top of the equilateral triangle spanned by that middle third.

Some generations of the Koch curve are displayed below.

Gen. 2

Gen. 3

Gen. 4

Gen. 8

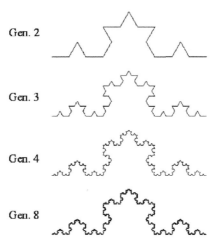

Figure 2.14: Koch curve generations

The limit is a fractal curve of the type that caused some anxiety among the mathematicians (Weierstrass and others) of the nineteenth century who were at work building the rigorous foundations of the calculus. This curve is an example of a one-to-one continuous mapping of the unit interval into the two-dimensional plane, which does not have a well-defined tangent line anywhere. In fact, the curve "seems" to be turning at an angle of 60° or 120° at all points.

The Koch *L*-system grows quite rapidly in terms of the number of line segments plotted. Since each line segment is replaced by four more at the next generation, the *n*-th generation has 4^n segments. For instance, generation 8 plotted above has 65536 segments.

2.5.1 Branching and bracketed L-systems

Many biological forms are "branched", "fragmented", or "cellular" in appearance and growth. To allow branching to occur in the turtle graphics interpretation of *L*-systems, we use the symbols [("push") and] ("pop") to enable the turtle to follow a branch for a time at the end of which it returns to the position where it started branching. We will begin with a simple branching where a main "trunk" shoots off one side branch.

    Angle 10
    Axiom F
    F=F[+F]F

This can be viewed as replacing each straight edge by the following configuration:

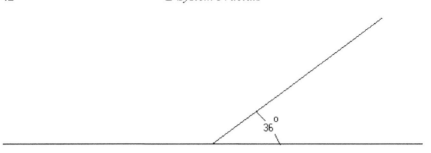

Figure 2.15: Simple branch

The change of angle and one step is taken with the pop and push operators. Scaling so that the main trunk remains the same size, we see the sequence of generations in the figure below.

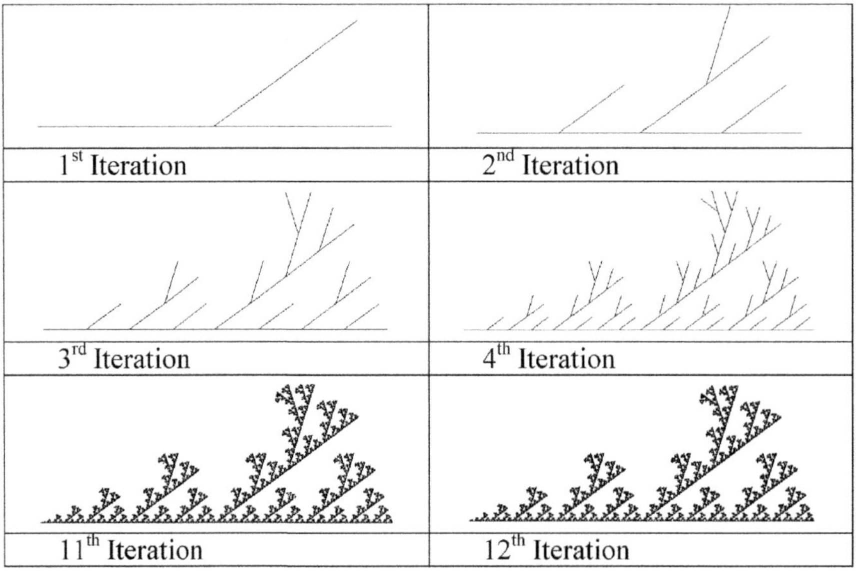

1st Iteration	2nd Iteration
3rd Iteration	4th Iteration
11th Iteration	12th Iteration

Figure 2.16: Generations of branching

To end this section, we give a picture of the BUSH (by Adrian Mariano) system. The model is single branching, slightly asymmetrical and with slightly curved branches.

Angle 16
Axiom ++++F
F=FF-[-F+F+F]+[+F-F-F]

The angle 360/16=22.5 degree allows for a gentle curving. The main trunk exists only before the branching, not afterwards. The four +'s in the axiom orient the structure vertically. The various generations of the L-system results in the following picture as shown in the figure 2.17.

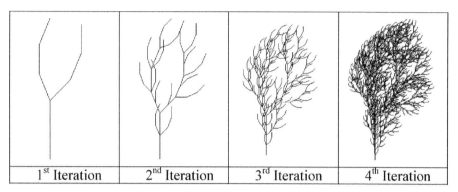

1st Iteration	2nd Iteration	3rd Iteration	4th Iteration

Figure 2.17: Bush *L*-system, generation

As we pass to later generations, the branches of the bush system cluster to give a three-dimensional appearance to the bush. True three-dimensional *L*-systems can be achieved by replacing + and - by a set of six three-dimensional rotations, one pair for each axis of rotation among the *x*-, *y*- and *z*-axes.

2.5.2 Famous L-systems of mathematical history

One of the great themes of nineteenth century mathematics was the collective effort of many mathematicians to put the methods of calculus and higher analysis on a firm rigorous foundation. The bane of modern calculus students, the $\in$-δ definition of limits, is one of the great highlights of this efforts. With the powerful new ideas about the exact notions of "limits" and "continuity", mathematicians were at last able to explore entirely new kinds of geometric objects.

Georg Cantor undertook a pioneering study of the nature of sets and real numbers in the last half of the nineteenth century. Although the idea of real numbers as infinite decimal expansions had existed for some time before Cantor, Cantor was the first to realize some of the surprising implications presented by this idea. Perhaps his most famous argument is the simple observation that given any enumeration of real numbers in decimal form, say:

$x_1 = n_1.a_{11}a_{12}a_{13}....$
$x_2 = n_2.a_{21}a_{22}a_{23}....$
$x_3 = n_3.a_{31}a_{32}a_{33}....$
....

This shift in mathematical philosophy, although it seems almost devoid of controversy today, is similar to the transformation that has taken place in dynamics over the past several decades. The predictable smoothly varying long-term behavior that was so highly prized before is now known to be the exception rather than the rule. The limiting geometry and dynamics are far more likely to be *fractal* or *chaotic* in nature.

Cantor's manipulation of decimal expansions reflects a perception of digit expansions as a symbolic dynamical system. The symbolic states are simply the digits 0 through 9. A decimal expansion is a listing of the states over each generation; the number itself is the behavior of the dynamical system. This idea suggests all kinds of curious things to consider. Cantor formulated a particularly interesting set based on restricting the "symbolic dynamical systems" of digit expansions. We need to make a slight variation from our usual system of expansion. Since there is nothing sacred about the use of the base 10, we can also work with expansions to other bases, particularly base 3. In base 3, there are exactly three digits 0, 1, and 2. A digit expansion of a number (called a *ternary* expansion) takes the usual form

$a_k a_{k-1} \ldots \ldots a_1 a_0 . a_{-1} a_{-2} \ldots \ldots \ldots \ldots \ldots \ldots \ldots$

where each digit a_i is 0, 1, or 2. The actual number is composed of powers of 3 (instead of powers of 10) as a series:

$a_k 3^k + a_{k-1} 3^{k-1} + \ldots . + a_1 3^1 + a_0 3^0 + a_{-1} 3^{-1} + a_{-2} 3^{-2} + \ldots \ldots$

Cantor defined a set C to consist of all numbers with a ternary expansion that used only 0's and 2's but no 1's. An expansion that ended with all 2's is allowed in C. The diagonal argument can be used to show that C is again uncountable, and therefore in that sense it has a great many points. However, as we shall see, it appears as a wispy smattering of points in line, with no length in any sense.

To "plot" the Cantor set, we shall consider only those points in the unit interval [0,1]. Considering the first digit, points beginning with 0.1…..do not occur in C, with two exceptions. The endpoints 0.10000….=0.02222…. and 0.20000….= 0.12222…. and have at least one possible expansion in all 0's and 2's and so belong to C. Thus, in trying to carve out C, we must first cut away the middle third of the interval (1/3,2/3). When we consider the second digit, we must also eliminate expansions beginning 0.01….and 0.21 and 0.21. These correspond to the middle thirds of the remaining intervals (1/9,2/9) and (7/9,8/9). This process of cutting out middle thirds continues indefinitely. The residue that's left over is the Cantor set.

This carries out the carving of the middle thirds. The sequence of the generations is shown below.

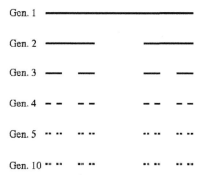

Figure 2.18 Generations of the Cantor *L*-system

The dust-like limit at the end is hardly impressive; yet, this kind of dusty distribution appears throughout nature, for instance, the clumpy distribution of matter in the universe. Mandelbrot calls this kind of geometric object *fractal dust*.

2.5.3 Self-similarity and scaling

Take a moment and review some of the pictures arising from the *L*-systems we have considered so far. The limiting geometrical object is an example of what is generally called in dynamical systems theory an *attractor*. Generally, this means some physical situation that the system in some sense converges to. For some symbolic dynamical systems, the attractors are the infinite sequences produced. For the *L*-systems coupled with turtle graphics, the attractor may be the limiting curve or shape. To arrive at the limiting shape, we must rescale the generations so that the "step size" d is decreased by a suitable factor from generation to generation.

When we view the limiting shape (or close approximations), the evolution and the rescaling are often visually apparent. Consider first the dragon curve. In this figure, we show how the limiting shape can be split into two congruent shapes.

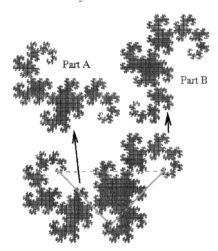

Figure 2.19: Splitting the dragon: The two halves of the dragon above are marked as Parts A and B.

The original isosceles right triangle is marked in the figure. The two smaller halves are each constructed on a segment of length $\sqrt{2}\,/\,2$ times the length of the segment of the original dragon. Also the two halves are congruent: one may lay one exactly on top of the other by a $90°$ rotation and a translation. We see that the dragon fractal is composed of two congruent pieces, each of which is scaled down by $\sqrt{2}\,/\,2$ from the original dragon. This is an example of *self-similarity*: the similarity of the whole object to smaller pieces of itself. To many of the graphic L-systems listed in FRACTINT and elsewhere we can associate two numbers: a "number of congruent pieces" that the shape may be divided into, and a "scaling factor" that scales each piece into the original object. For the dragon fractal, the number of pieces is 2 and the scaling factor is $\sqrt{2}$ these numbers can partly be deduced from the L-system rules, although a universal method for doing this is not known. For the dragon, we see in the L-system production rules that each segment reproduces two smaller segments of length $\sqrt{2}\,/\,2$ times the original. As a second example, consider the Koch curve, which we show below splitting into similar parts. The production rule converts each segment into 4 new ones of length equal to 1/3 the original length. Thus, it is not too surprising to see four similar pieces each one-third the size of the original Koch curve.

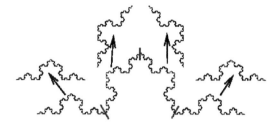

Figure 2.20 Splitting the Koch curve

Self-similarity is a key concept in the theory of dynamical systems and the resulting fractal geometry.

2.6 Summary

The review of various literatures in this chapter reveals the idea of generating fractals using L-System approach. Although many methods have been used to generate using the L-System and IFS approach the combination of two L-System models, one for the base Koch curve and that of L-System for recursive mathematical formulae have been shown in this book to generate a class of amazing linear fractals. A critical review in the subject of parallel grammar has been made to visualize its application for designing a complete compiler based on parallel grammar. As this chapter highlights the basic background for generating fractal figures using L-System approach, a number of figures have been generated and shown in the subsequent chapters. A class of hybrid fractals have been generated by the authors to illustrate the use of L-System for generating fractal figures.

Chapter 3

Interactive Generation of Fractal Images

The formulation of mathematical equation for generating fractal figures is emphasized in this chapter. A discussion about the iterated function for self-similar pictures or images is made. As Iterated Function Systems or IFS plays an important role in generating figures the same figure can be downloaded or uploaded with less downtime in networking environment with only the mathematical equations which can take less memory. A brief introduction of IFS is presented in relationship with computer graphics particularly, fractals [37,57,59,61,88,123].

3.1 IFS and Fractals

Some quantifiable relationship does exist between fractal images and mathematics. A few fractal figures using IFS are discussed as the base concept. A software has beed developed to generate fractal figures and produce IFS codes that can be used in conventional programming to generate fractal figures. Barnsley tried to solve the inverse problem [88,89] i.e., given figures how to generate equations both for such similar and dissimilar figures.

Unlike familiar geometrical shapes, which have integer values for topological dimension (simply speaking, a point has zero dimension, a line dimension 1, a square and sphere dimensions 2 and 3 respectively), the dimensions of fractal shapes can take fractional values. This is known as Hausdorff-Besicovitch (HB) or fractal dimension. For a fractal, the HB dimension is strictly greater than the topological dimension. Moreover a fractal curve can be continuous everywhere, but nowhere it is differentiable. A third interesting characteristics of fractals is that they are self-similar, i.e., small parts of the fractal appear to be condensed versions of the whole. This introduces a kind of order in a seemingly irregular pattern. Each of these above properties can be seen in familiar mathematical curves or functions. Therefore the introduction of a new geometry to describe families of highly complex irregular shapes is justified.

IFS (iterated function system) is another way of generating fractals. It is based on taking a point or a figure and substituting it with several other identical ones. For example, there is a very simple method for generating the Sierpinski triangle. One can start with an equilateral triangle and substitute it with three smaller triangles. By iterating this process you substitute each of those three triangles with three even smaller triangles and continue a large number of times. Mathematically, substituting a shape with another one is called a geometric transformation. The above example two kinds of transformations: translation (movement of triangles) and dilation (changing the size of the triangles). The third kind of transformation is rotation. It can be used to create fractals in which the self-similar parts are located at different angles. For example, to create a realistic fractal model of a tree you will need rotation for the branches. Other types of transformation, such as reflection and inversion can also be used to create a great variety of fractals. IFS make it relatively easy to create algorithms for drawing fractals. For most 2-D fractals, all you have to store is the list of all transformations with 6 parameters each: horizontal movement, vertical movement, rotation of the figure's vertical axis, rotation of the figure's horizontal axis, the stretching of the figure's vertical axis, the stretching of the figure's horizontal axis.

3.2 Generation of Fractals

These following figures (Fig 3.1 and Fig.3.2) are some of the common classical fractals.

Fig. 3.1: Classical Fractal

Fig. 3.2: Classical Fractal

Closely related to self-similarity is another key feature: Since the whole structure has gaps, there will be no part of the fractal, which really fills out its area. So, the whole thing has no area. It is not something 2-dimensional. Well is it 1-dimensional then? A one-dimensional object is similar to a line and has a length. But if you look at the second fractal, it seems to be made up of infinitely many line segments on a smaller and smaller scale. As the length bits do not sum up to something finite, you cannot assign a length to the whole fractal. So its dimension is between 1 and 2, it has a **fractal dimension**, which is the motivation for the name. This is only a very rough explanation.

3.2.1   Multi Lens Copy Machines

The idea of a multiple lens copy machine [57] helps to understand IFS-fractals, all fractals actually. The idea to use this illustration can be taken from the book "Fractals for the Classroom" by Peitgen, Saupe, Richter. Suppose the special copy machine has 3 lenses that can make 3 different copies of the original on the same sheet like figure 3.4.

Fig. 3.3: Initial Image

The initial image is figure 3.3. It is reduced by half and pasted in three different places by the three different lenses as shown in fiure 3.4. This is called the **blueprint** of the fractal to be generated. These lenses don't distort or flip the original, as can be seen by the way the letter L is copied. Now this copy is taken as the original and copied again (Fig. 3.5 & Fig. 3.6):

Fig. 3.4: First Blue Print

Afterwards, the process is repeated again and again a number of times.

L-System Fractals

Fig. 3.5: Next Iteration

Fig. 3.6: Again the next iteration

What will happen as we keep going (Fig. 3.5, Fig. 3.6 & so on) ? Here's a picture of the 10th copy in figure 3.7.

Fig. 3.7: 10th iteration

The copy and the original are virtually indistinguishable. There is an idealized object, which the copies approximate: Its copy will not differ from itself. It means that you can find copies of it in itself on -all- kinds of smaller scales. This thing is a fractal and the name of this particular one is Sierpinski Triangle or Sierpinski Gasket. It may be seen that all the self-similar copies of the triangle are there within itself. Now, as an example the Sierpinski triangle is good because it is not too confusing, but its picture doesn't look too exciting. The next picture shows the first copy of a page for a very famous copy machine. The blueprint has the L's in it so one can see whether a page is rotated or flipped.

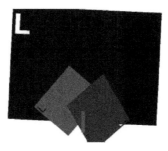

Fig. 3.8: Blueprint containing a rotated image

The little line in the middle of the bottom in figure 3.8 may be seen. The blueprint should have every information in it to be able to figure out the final fractal. Once iterated several times we will get the final fractal as shown in figure 3.9.

Fig. 3.9: Bransley's Fern

This is Michael Barnsley's well-known fern. The fractal is overplayed with its blueprint so we can see that indeed all copies look like the whole page and that the fractal is being copied onto itself. Do we have to start with the whole page being black in order to generate the fractal? This series of pictures starts out with a fractal plant and then uses the Sierpinski copy machine a number of times. What can be learned from this is that the only thing that matters for the final fractal shape is the copy machine design and not the structure on the first page.

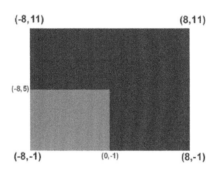

Fig. 3.10: Blueprint for a copy machine design

Here's how lens 1 of the Sierpinski machine (the lens that goes to the left bottom part) works: We assume the original page to have its edges at the points as shown in figure 3.15 with edges at (-8, -1), (8, -1), (-8,11), (8,11). This is a standard IFS convention, based on full screen graphics, we guess. Now if (x,y) is a point on the original, then the copied point will have coordinates (x',y') given by

x' = 0.5 x + 0 y + (-4)
y' = 0 x + 0.5 y + (-0.5)

We can check this when we plug in some points. It may be noted that the lens scales down both the x and y coordinates by a factor 0.5 and then shifts x coordinate by –4 and y by -0.5. This takes the copy to the lower left edge. Now we can see 6 numbers, which encode the lens. It is common to list them in this order:

0.5 0 0 0.5 -4 -0.5

More general copy machine lenses are made up by a rule

x' = a x + b y + e
y' = c x + d y + f

and the numbers

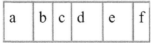

a	b	c	d	e	f

are the codes of the lens. These rules are called **Affine Transformations**. All affine transformations are combinations of shifts, rotations, scalings and shears, which means, we can still see the similarity between the original and the copy. The Sierpinski Copy Machine has 3 lenses, so there are 3 times 6 = 18 numbers that encode the machine. They are commonly listed in 1 row for each lens. So this is the IFS-code for the Sierpinski triangle:

0.5	0	0	0.5	-4	-0.5
0.5	0	0	0.5	4	-0.5
0.5	0	0	0.5	0	5.5

Table 3.1: IFS Codes for Sierpinski triangle

If we can think about how those lenses work, neither is the code too difficult to figure out, nor is it too difficult to imagine what this fractal looks like in the end. This changes with less straightforward lens systems, as, in the example of the fern.

3.3 Computer Implementation

3.3.1   The Random Algorithm

Following the copy machine idea, computers could faithfully copy pixel by pixel according to the rules of the IFS, and repeat this until the fractal shape is sufficiently well approximated. Now this is of course very slow, besides being very unprecise, as pixels have a thickness, and real numbers have not. A better algorithm does the following:
Start out with any point (x,y) in the window.
Throw a dice between the lenses.
Copy (x,y) by the picked lens to a new (x,y) and plot that one (now scaling real numbers to screen pixels, of course).
Repeat steps 2 and 3. If you disregard the first 50 or so points, the rest 50000 or so trace out the fractal pretty faithfully. Here's just an idea: You have seen how the copies more and more home in on the final shape. A randomly copied point would do the same, it has to get closer and closer to the fractal. But why should a tail end of such a point sequence necessarily approximate the complete fractal? In theory, not just any initial point works, but with probability 1 you pick one that does, because, well -here it gets a little difficult, otherwise the fractal would be so thin, you would not be interested in it, because you couldn't see it in the first place. Basic implementation of the random algorithm is rather easy. You should try it in your favorite programming language with the fern:

0.85	0.04	-0.04	0.85	0	1.6
0.2	-0.26	0.23	0.22	0	1.6
-0.15	0.28	0.26	0.24	0	0.44
0	0	0	0.16	0	0

Table 3.2: IFS codes for fern

You would probably see that the tip of the fern is a bit thin. This is because the large lens, which "moves the leaves up the stem" needs to get a little boost. This can be implemented by assigning different probabilities to the lenses, according to the area their images cover: Probability = Area / TotalArea. You can also assign colors to the lenses, and use a coloring scheme, which brightens or darkens points according to the frequency by which they are hit. Adding all this information to the IFS code, each lens could have a code of eight numbers:
a b c d e f probability color.

3.4 Designing Fractals

The idea is simple: Encode the blueprint by triangles. Say you give yourself a reference triangle, which is half of the original page, and then you draw another triangle,

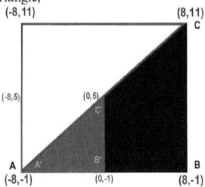

Fig. 3.11: Triangle with coordinates

which would be the copy of the first lens, as here for the Sierpinski triangle as shown in figure 3.11. Then read off the coordinates of the points A', B', and C'. Those are 6 numbers , which you can use in term to figure out the 6 numbers for your IFS-code. It "only" involves solving 6 equations with 6 unknowns. a, b, c, d, e, f are the unknowns. For x, y you plug in the coordinates of A, for x', y' the ones for A'. That gives the first 2 equations. The rest of the triangle edges give the other 4 equations. This algorithm works with arbitrary copy triangles. The above-mentioned area of mathematics known as Linear Algebra deals with the efficient solution of such equations (though the solution formulae are actually not so difficult to figure out by hand). But let's say we leave solving the equations to the computer once we've drawn the triangles for our lenses. Note, that it's not important what we choose as our reference triangle. The only thing that matters is, that the other one is a copy of the reference triangle. This idea leads to an interactive graphical fractal designer, which lets you design your lenses graphically by adding and manipulating their triangles. As a feedback, a random algorithm can display the output of the final fractal, with the fractal reflecting the changes in the triangles. Randomly chosen triangles (or IFS codes) usually give blobs or haystacks or needles. To give you an idea of how to design a fractal, here is a picture of a fractal together with the triangles that design it. The one labeled A0, B0, C0 is the reference triangle. It is supposed to be the reference for the whole picture. See how the others contain copies of what is in the reference triangle, and how all parts of the fractal are accounted for by the

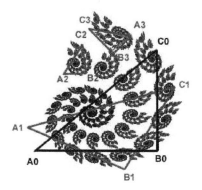

Fig. 3.12: Reference Triangle

smaller triangles. Note that for fractal coverage you have to extrapolate the reference triangle to the whole page, and the copies accordingly. You can experiment with loading some of the predefined fractals and see what happens when the triangles are changed. It should be noted that fractals can be designed to resemble leaves or other roughly self similar real objects, by copying the object or a picture of it onto a transparency, then putting it into a reference triangle and drawing smaller copy triangles, just as in the above picture. The transparency you can then use to enter the triangles into the computer. Since no real thing is really self similar (except for this roman cauliflower at the grocery store), there are a little fantasy and experimenting involved in this. One of the many ways of describing the affine transformations for iterated functions systems (IFS) is as follows:

$x = r\cos(\text{theta}) \, x + s \sin(\text{phi}) \, y + h \quad y = -r \sin(\text{theta}) \, x + s \cos(\text{phi}) \, y + k$

Generally there are a number of these mappings (different values of r, s, theta, phi, h, k) which are chosen at random at each iteration.

3.4.1   How does the program work

The fundamental iterative process involves replacing rectangles with a series of rectangles called the generator. The rectangles are replaced by a suitably scaled, translated and rotated version of the generator. For example consider the generator below as shown in figure 3.13.

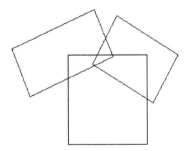

Fig. 3.13 The generator

It consists of three rectangles, each with its own center, dimensions and rotation angle. The initial conditions usually consist of a single square, the first iteration then consists of replacing this square by a suitably positioned, scaled and rotated version of the above generator. The result would just be the generator above. The next iteration involves replacing each of the rectangles in the current system by suitable positioned, scaled, and rotated versions of the generator resulting in the following figure as shown in fig.3.13. The next iteration replaces each rectangle above again by the initial generator as shown below and so on. During three iterations as seen in fig. 3.14, Fig. 3.15 & Fig. 3.16, we get our final fractal.

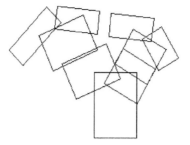

Fig. 3.14: First Iteration

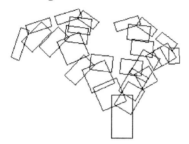

Fig. 3.15: Second Iteration

Fig. 3.16: Third Iteration

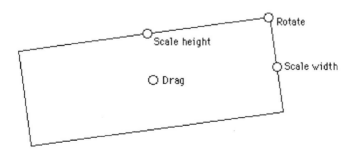

Fig. 3.17 The rectangle generator

IFS allow the user to explore the result of applying different generators. The generator rectangles can be created, shifted, scaled, and rotated by the handles on each rectangle shown above in fig. 3.17.

Fig. 3.18: Leaf   Fig. 3.19: Leaf

The user may iterate forward or in reverse, colour the borders, interior or background. A library of interesting generators is provided for initial experimentation. Generators created by users may be saved in files and recalled at a later time. Images so created may be copied from IFS and pasted into a more standard drawing package for incorporation into a document and printing. The above figures (Fig.3.18 & Fig. 3.19) show a maple leaf generated using the IFS program.

3.5 Software Package

A software package has been developed using VB (Visual Basic) for interactive creation of fractal images by visual manipulation of blueprints. A representative list of figures generated by using this package is presented in the illustration. The inverse problem of determining the parameters of the IFS code from the blueprints is also solved through the package.

3.5.1   Background

Iterated Function Systems (IFS) plays an important role in generating fractal figures [123]. A software package is developed using Visual Basic (VB), which can create the blueprint, generate the corresponding fractal pattern and extract the IFS codes that can later be used to reproduce the original fractal figure.   An attempt has also been made to solve the inverse problem, i.e., estimating parameters of the affine transformation that serve as the IFS code for the given self-similar image. These transformations comprise of combinations of shifts, rotations, scaling and shears.

For example, Table 3.1 presents a matrix of six transformation parameters for the three replicas in the blueprint needed to generate, through an iterative algorithm, a fractal figure called the Spierpinski triangle shown in Fig 3.25

0.5	0	0	0.5	-4	-0.5	
0.5	0	0	0.5	4	-0.5	
.5	0	0	0.5	0	5.5	

Table 3.3: IFS codes for Sierpinski Triangle   Fig. 3.20: Sierpinski Gasket

3.5.2   Computer Implementation

The program works in two parts. In the user defined first part, the coordinates of the vertices of an initial polygon are stored through a mouse click. On this figure one can add polygons, relocate the polygons through dragging and arrange them in a collage interactively to create a variety of IFS blueprints. The resulting fractal obtained as the limiting case of the iterative algorithm is displayed simultaneously in a window. Using the coordinates of the vertices of the polygons defining the blueprint, the parameter table is calculated through the solution of a set of simultaneous equations by Gauss elimination method. Also in this package one can simultaneously manipulate the blueprint as well as figures in two windows in the same screen.

In second part, i.e., automatic figure generation part, transformations defined by the given parameter values are iteratively applied to an arbitrary initial figure to generate the corresponding fractal pattern [10].

3.5.3   Sample Output

The table 3.4 shows some of the blueprints and the corresponding fractal figures generated by using these codes. The colour images of the generated fractals are shown in the colour section at the end of the book. The respective parameter matrices can be calculated by using the package developed.

BLUE PRINT	GENERATED FIGURES
The Sun (Blue print)	The Sun
Leaves (Blue print)	Leaves
The Goddess Subhadra (Blue print)	The Goddess Subhadra

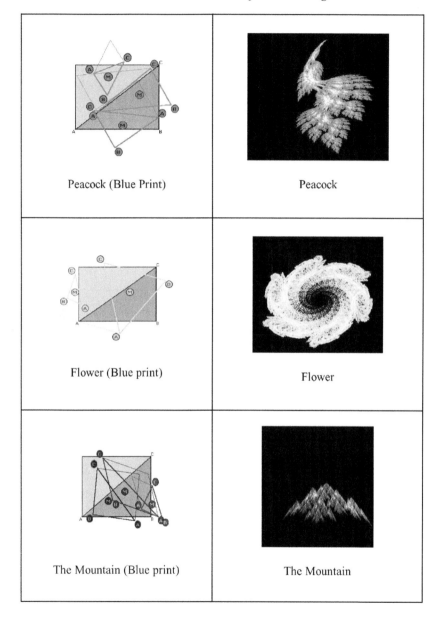

Peacock (Blue Print)	Peacock
Flower (Blue print)	Flower
The Mountain (Blue print)	The Mountain

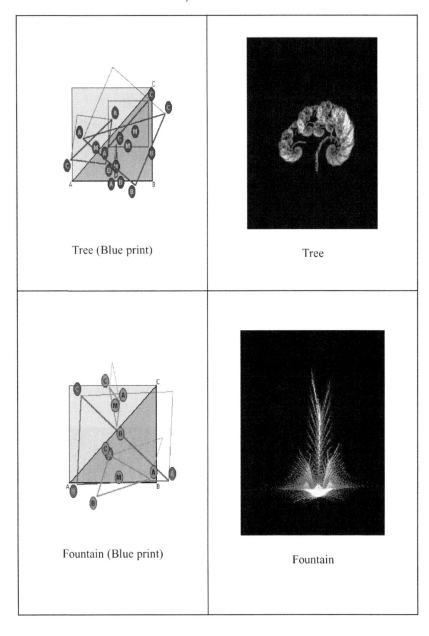

Tree (Blue print)

Tree

Fountain (Blue print)

Fountain

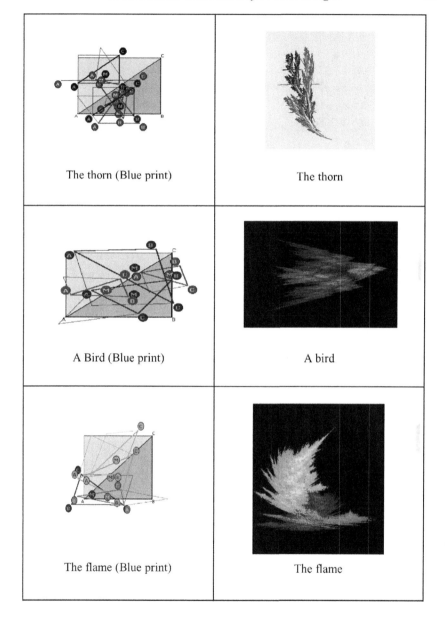

The thorn (Blue print)	The thorn
A Bird (Blue print)	A bird
The flame (Blue print)	The flame

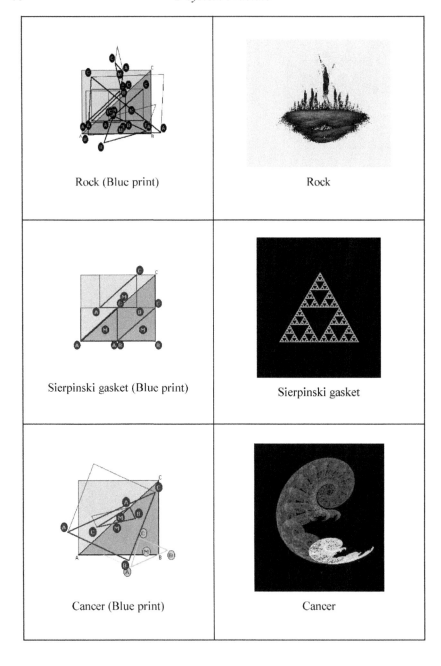

Rock (Blue print)

Rock

Sierpinski gasket (Blue print)

Sierpinski gasket

Cancer (Blue print)

Cancer

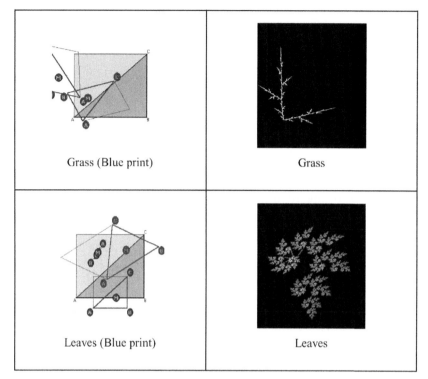

| Grass (Blue print) | Grass |
| Leaves (Blue print) | Leaves |

Table 3.4 : Figures generated by using blue prints

3.6 Mathematical Expression of IFS

An iterated function system consists of a set of maps $\{w_i\}_{i=1}^{N}$ from $\mathbf{R}^n$ into itself. If the maps of an IFS are contractive, each IFS includes a single, compact, nonempty set $A \subset \mathbf{R}^n$, called its attractor defined as the union of images of itself under the IFS maps:

$$A = \bigcup_{i=1}^{N} w_i(A)$$

The general form for a contractive affine transformation is:

$$w_i(x) = \begin{bmatrix} a_{11}a_{12} \\ a_{21}a_{22} \end{bmatrix} x + \begin{bmatrix} s_1 \\ s_2 \end{bmatrix} = A_i x + b_i$$

where $|det(A_i)| < 1$ and x in $\mathbf{R}^2$. Because fractal images have self-similarity, we may find an IFS code consisting of affine transformations w_i, i =1,2,.....,N such that:

$$F = \bigcup_{i=1}^{N} W_i(F)$$

where F is a fractal image and $w_i(F)$ is a sub image given by:
$w_i(F) = \{ z \in R^2 \mid z = A_i x + b_i \text{ for } x \in F \}$.
This is nothing but the famous collage theorem

The Hutchinson operator **w** is a convenient shorthand notation

$$\mathbf{w}(\cdot) = \bigcup_{i=1}^{N} w_i(\cdot)$$

that allows us to simplify the definition of an IFS attractor as

$A = w(A)$

Furthermore, the attractor A gets its name from the property that, given any initial nonempty bounded set $B \subset \mathbf{R}^n$, we have

$$A = \lim_{j \to \infty} \mathbf{w}^{0i}(B)$$

where w^{oi} denotes the I-fold composition of **w** (i.e $\mathbf{w}^{oi} = \mathbf{w} \text{ o } \mathbf{w}^{oi-1}$).

3.6.1 RIFS

Likewise, a recurrent$_N$iterated function system consists of a set of affine transformations $\{w_i\}^{i=1}$ and a directed graph G. Each edge $<i, j> \in G$ indicates that the composition $w_j \text{ o } w_i$ is allowed. If a RIFS consists of contractive maps, there exists a single compact nonempty attractor $A \subset R_n$, defined as a collection of possibly over lapping partitions $A_j \subset R_n$:

$$A = \bigcup_{j=1}^{N} A_j$$

which are each the union of an image of other partitions (including possibly itself):

$$A_j = \bigcup_{(i,j) \in G} w_j(A_i)$$

We can keep the components of this partitioning separate by denoting the attractor $A = A_1 \cup A_2 \cup \cdots \cup A_N$ as a vector of sets $A = (A_1, A_2, \ldots, A_N) \in (\mathbf{R}^n)^N$.

The domain of the RIFS maps extends to set vectors $w_j : (\mathbf{R}^n)^N \to \mathbf{R}^n$, which are defined

$$w_j(\boldsymbol{A}) = \bigcup_{(i,j)\in E} w_j(A_i)$$

Using these maps, the recurrent Hutchinson operator $w:(\mathbf{R}^n)^N \to (\mathbf{R}^n)^N$ is defined on set vectors as

$w(A) = (w_1(A), w_2(A), \ldots, w_N(A))$

The attractor, consisting of the components defined earlier, is now more concisely defined in set-vector notation as

$A = w(A)$

to better resemble the definition of IFS attractor. Moreover, given any set vector $\boldsymbol{B} = (B, B, \ldots, B) \subset (\mathbf{R}^n)^N$ consisting of bounded nonempty sets $B \subset \mathbf{R}^n$, then

$$A = \lim_{j\to\infty} w(\boldsymbol{B})$$

So, given any initial nonempty bounded set $B \subset \mathbf{R}^n$, iterating w on the set vector $\{B, B,...,B\} \in (\mathbf{R}^n)^N$ creates a sequence converging to the set vector A.

If each partition is the union of images of every partition including itself, then the graph G is complete and the RIFS is simply an IFS. Hence, every IFS is an RIFS.

3.6.2 Modified MRCM

A modified MRCM has been proposed by Bisoi & Mishra [5,6], where the reduction of the image is done by averaging out the intensities of neighboring pixels rather than simple scaling down of the outline of the image.

While obtaining the collage of the reduced image massic transformation such as color reversal is applied to selected copies of the reduced image.

This coping machine again takes the output image as its input and repeats the process as it gives a series of fractal patterns.

Let us consider an example with an adjustment factor AJ = $(J_i,$ i=1,2,.....,N), which represents n massic transformations that will make gray scale adjustment through color reversal and thus $Z_i = W_i \circ J_i$ are the adjusted IFS codes such that

$$F = \bigcup_{i=1}^{N} Z_i(F)$$

3.7 Summary

One is amazed by the complexity, beauty and diversity of the generated gallery of fractals. Many of the resulting patterns resemble natural objects. It appears that to understand natural shapes, one must understand the IFS codes thoroughly.

In networking environment, at the time of downloading any self-similar figure from a network server, only the parameter matrix in text format needs to be transmitted which can be used automatically in client machine to regenerate the fractal image either in online or offline mode, which may help us to save downloading time. Transmitting images in full binary or compressed binary form, in contrast, would require more memory as well as time. Research is under progress to represent an arbitrary given image by its fractal code.

This concept is meant for researchers not only to think about generating linear IFS code for various similar figures but also for 3D nonlinear figures. IFS concept is being used in many practical fields like weather forecasting, medical sciences and earthquake forecasting and natural figures generations. Furthermore, the field of fractal figures using IFS concept is far from being exhausted since there are many directions that have not yet been fully investigated (e.g., the use of non-affine transformation, combination of fractal coding with other techniques and video frames etc.). We hope the reader will understand the technique to be able to perceive the application of IFS coding for generation of similar figures in computer graphics.

Chapter 4

Generation of a Class of Hybrid Fractals

A new method for generating a class of fractals associating the axioms and productions of classical fractal, like Koch curve and its multiple variations represented in L-system, with that of the mathematical series represented in L-System is presented in this chapter. The turtle graphic interpretation has been used and infinite types of fractals have been generated. Various types of fractals are easily represented by L-Systems and generated by means of a given graphic interpretation. However, one must distinguish clearly between the L-System itself and the graphic interpretation that generates the fractal.

The possible combinations of axiom and rules used in Koch curve with that of axiom and rules for the recursive mathematical formulae in L-system formats have been tested with turtle interpreted software for generating a new hybrid class of fractals. As any non-graphic symbol has no impact on turtle interpretations, English alphabets like A, B, C..etc, have been used as variables required for representing the mathematical sequences represented in L-System.

4.1   Background

4.1.1   Parallel grammar: A critical review

From an abstract level, parallel grammars (and automata) try to model phenomena encountered in parallel processing. Since in grammatical mechanisms for generating formal languages the basic and simplest processing step is that of a replacement of one symbol (or a string of contiguous symbols in the non-context-free case, but this can be seen as a special case of the context-free case if we consider parallel derivations restricted to adjacent symbols) by a string of symbols. The most straightforward way to incorporate or model parallelism here is certainly to consider the possibility of doing a number of derivation sub-steps in parallel; hence defining the notion of a *parallel derivation step*. One obvious possibility here is to allow (and enforce) the parallel execution of a context-free derivation step in all possible places. This basically leads to the theory of Lindenmayer systems [12,13,14,42]. For

physical reasons (and often also biologically motivated), it is unreasonable to assume that basic processing units (cells) located in the midst
of a larger computing facility (body) may grow arbitrary (by "splitting"
the cell due to some rule of the form $A \rightarrow BC$). This idea led, on the one
hand, to the theory of cellular automata [73]. In this context, array
grammars are mentioned, which adhere to similar growth restriction,
although they are not necessarily parallel in nature, but as often the
case in formal language theory, different concepts of rewriting; here the
ideas of array grammars and of parallelism in rewriting can be fruitfully
combined [30,52,111,114]. Even more interesting than the usually considered one-dimensional case are the higher-dimensional cases, where
the restrictions of growth in the innermost cells of the body become
even more obvious. When comparing the growth patterns, i. e., the
development of the length of the cellular strings as a function of time,
which is typically found within (certain types of simple) Lindenmayer
systems, we find that they mostly show polynomial or exponential
growth. On the other hand, in nature, we never encounter unlimited
exponential growth; this sort of growth pattern can only be observed
over a limited time (even with cancer cells or other cells living as
"parasites"), since at some point of time the host will die and this will
immediately stop any growth process. The most "natural" way to
model this phenomenon is to incorporate environmental factors, which
mostly determine the limitation of growth, within the formal language
model itself. In this way, we arrive at several forms of limited parallel
rewriting as discussed in the forms of ecogrammar systems [34].

When discussing partial parallel rewriting, we could consider the
classical forms of sequential rewriting as a very special extreme case.
Coming from this end of the possible spectrum of parallel rewriting, it
is quite natural to consider also the parallel replacement of only "a few"
symbols as one rewriting step. As usual, both "grammatical" and
"automata-based" models can be considered here. Due to the multitude
of possible replacement strategies, we mention only a few of them in
the following:

Equipping finite automata with more than one head can enhance
the power of finite automata beyond accepting regular languages
[84,56]. This idea is closely related to certain forms of parallel
grammars basically leading to a form of absolutely parallel (array)
grammar.

In a grammatical mechanism where the replacement of terminal
symbols is not permitted, allowing a grammar to produce only
sentential forms containing a bounded number of nonterminal
occurrences is a severe restriction basically known as "finite index

restriction" Infact, regulated grammars of finite index (which in case of most regulated rewriting mechanisms turn out to be of equal generative power, for a recent account incorporating lots of references) can be equivalently "interpreted" as doing one restricted parallel derivation step involving a maximum of k nonterminals, where k is the finite index of the grammar and all (l) nonterminals of a sentential form w must be replaced by a rule of the form $(A_1,\dots,A_l) \to (w_i,\dots,w_l)$, and the nonterminals A_1 through A_l appear in that order in w. These *absolutely parallel grammars* introduced by V. Rajlich [132] generalize towards *scattered context grammars* if we do not insist on replacing all symbols of a sentential form in parallel, but can select some of them (obeying the order imposed by the rule). A related notion is that of simple matrix grammars [38,96,117].

- Taking into account complexity restrictions, it is also interesting to study Turing machines having multiple heads .
- Instead of allowing or even enforcing that *all* occurrences of symbols of a sentential form are replaced in one step, only a certain limited number of replacements per step is allowed. This leads to the theory of (uniformly) limited Lindenmayer systems, which we discuss further in some detail below.
- Allowing finite automata or regular grammars to work in parallel (in a possibly synchronized fashion) and combining the results in the end is one of the historic roots leading towards what is now known as parallel communicating grammar systems. Besides allowing combination of results in the end, certain forms of communicating the results have been discussed [34].
- It is possible to consider a combination of different forms of language models of parallel computation.

For example, D. Watjen recently investigated parallel communicating limited and uniformly limited 0L systems [33], while Gy. Vaszil studied parallel communicating Lindenmayer systems in a series of papers [47,48]. Another more classical example is the so-called Russian parallel grammars, which combine features from sequential and (Indian) parallel rewriting [92].

4.1.2 Rules for biological phenomenon

Looking closer at modeling biological phenomena, it is quite obvious how certain cases often distinguished in formal language theory actually reflect different biological phenomena:

Erasing production: a rule of the form $A \to \varepsilon$ models the death of a cell (which has been in state A before its disappearance)

Chain rule: a rule of the form $A \to B$ reflects a change of state of the corresponding cell

Growing rule: a rule of the form $A \to BC$ models the split of a cell previously being in state A, having two "children" being in state B and C, respectively

Context: a rule of the form $ABA' \to ACA'$ shows that the context-free rule $B \to C$ is only applicable within an appropriate left and right context; which aren't morphisms any more from the strict algebraic perspective

Pure grammars do not differentiate between terminal and nonterminal symbols, so that all sentential forms generatable from the grammar's axiom are put into the generated language; this notion is well-motivated biologically since the alphabet symbols in L systems should denote certain "states" of cells, and it is not reasonable to consider only organisms consisting of cells being in peculiar states. But there are more things, which find rather straightforward modelisations in the theory of L-Systems, even though this does not reflect "usual" concepts in formal language theory. Stated positively, it shows (again) how ideas from other areas (like biology) led to interesting theoretical concepts in language theory. A few examples are mentioned below:

Offsprings: Even some multi-cellular organisms do not (only) rely on sexual reproduction, but they can "split" and produce offsprings in this way. (Sometimes, this sort of reproduction is only by accident, as exemplified by some worms, which (may) develop, into two separate organisms when split by force.) This phenomenon is called *fragmentation* [82].

Grown-up: Higher organisms go through a couple of stages until they reach full maturity. In an idealised interpretation, being grown-up or adult means that the overall "body" does not change anymore, although "local changes" due to the death of some cells, which is compensated by the birth of others, are possible. Related to this are the "death word languages" considered by M. Kudlek [90].

Interaction with neighbouring cells: For biology, organisms consisting of cells, which are not interacting with each other, seem to be quite "unnatural". Possibly, one would not consider such a hypothetical collection of cells as one organism on its own. Hence, it is doubtful whether all the studies undertaken on basically context-free partial parallel grammars can be meaningfully re-interpreted in biological terms. In fact, in the case of Lindenmayer systems, the incorporation of interaction was done in the very first paper, and interactionless systems were considered only later. In the case of limited Lindenmayer systems for example, interactionless systems were considered first [31] and only recently limited systems with interactions were investigated [33].

If interaction with neighbouring cells is incorporated in mathematical models and formalisms, then the question arises what neighborhood actually means, especially in the multidimensional case. In the theory of cellular automata, several notions of neighbourhoods were studied.

Signals: Closely related to the notion of neighbourhood is the idea of using this neighbourhood in order to transport information from one part of the body to another one. Information flow techniques (or *signals*) are the basis of various algorithms for cellular automata [118]. Especially with parameterized L systems, this issue arose also in L system models.

Changing environmental conditions: The change of growth patterns in organisms can have various reasons, one of them being the change of the environmental conditions which are responsible, e. g., for the annual rings observable in cut trees. This has led to the introduction of so-called *tables* in Lindenmayer systems, where each table is representing a certain special environment [40].

Non-observable internal changes: Often, it is not possible to directly observe what is going on in living organisms. The observable changes can be due to various reasons.

In the more abstract model of Lindenmayer systems, this means that we are not dealing with the language of words generated by an L system, but rather by codings of these words, where a coding is a "renaming" homomorphism [39].

4.1.3 Some definitions and examples

L systems (without interaction) are defined below and several variants
are discussed.

An *interactionless Lindenmayer system with tables,* for short a *TOL
system,* is given by a triple $G = (\sum, H, \omega)$, where the components fulfill
the following requirements:

-   $\sum$ is the alphabet.

-   *H* is a finite set of finite substitutions $H = \{h_1, \ldots
    , h_l\}$, i. e., $h_i : a \rightarrow \{w_1, \ldots, \quad W_{n_{i,a}}\}$, which means that each h_i
    can be represented by a list of context-free rules $a \rightarrow \omega$ such
    that $a \in \sum$, $\omega \in \sum^*$; this list for h_i should satisfy that each
    symbol of $\sum$ appears as the left side of some rule in *hi.*

-   $\omega \in \sum^*$ is the axiom.

Some special cases are:

-   $t = 1$: we have a *0L system.*
-   $\forall 1 \leq i \leq t \quad \forall_a \in \sum: \quad n_{i,a} = 1;$ *deterministic* T0L
    systems, or DT0L for short; in other words, each h_i is a
    homomorphism.

G defines a derivation relation $\Rightarrow$ by $x \Rightarrow y$ iff $y \in h_i(x)$ for
some $1 \leq i \leq t$, where we now interpret h_i as substitution mapping.
The language generated by *G* is L(G) = $\{\omega \in \sum^* \mid \omega \Rightarrow^* \omega\}$ with $\Rightarrow^*$
denoting, as usual, the reflexive and transitive closure of $\Rightarrow$.

Possible variants are the following ones:

-   Given some "terminal alphabet" Δ, one might consider
    the *extended language* $E(G, \Delta) = L(G) \cap \Delta^*$.

-   Given some coding function $c: \sum \rightarrow \Delta \longrightarrow A$, the
    language $c(L(G)) = \{c(w) \mid \omega \in L(G)\}$ can be investigated.

-   $A(G) = \{\omega \in L(G) \mid \{w\} = \{u \mid \omega \Rightarrow u\}\}$ is the
    adult language generated by G.

The corresponding language classes will be denoted by ET0L,
CT0L, and AT0L, respectively. Again, variants like "A0L" denoting the
adult languages definable by 0L systems can be considered.

Let us explain these variants by means of an example:

Example 1. Consider the language

$L=\{ d^n\, b^n\, c^n \mid n \geq 1\}$

$A \rightarrow AA', A' \rightarrow A'$
$B \rightarrow BB', B' \rightarrow B'$
$C \rightarrow CC', C' \rightarrow C'$
are the rules of a 0L system with axiom ABC which, together with
the coding $A, A' \rightarrow a$, B, $B'\; b$, and $C, C' \rightarrow c$ describes L. More
precisely, for the n-step derivation, we can easily see that
$ABC =>^n A(A')^n B(B')^n C(C')^n$
from which the claim follows.

Taking these same rules plus the rules
$A \rightarrow a, \qquad A \rightarrow a, a \rightarrow F$
$B \rightarrow b, B' \rightarrow b, b \rightarrow F$
$C \rightarrow c, C' \rightarrow c, c \rightarrow F$
$F \rightarrow F$
result in an E0L system (with axiom ABC) which also generates L
when taking $\{a, b, c\}$ as terminal alphabet.

Bharat (T0B) systems: $x \Rightarrow y$ iff $\exists 1 \leq i \leq t$ $\exists a \in \sum \Rightarrow$ such that all
occurrences of a in x are replaced by some word in $h_i(a)$ to obtain y
from x;

k-limited (kIT0L) systems: $x \Rightarrow y$ iff $\exists 1 \leq i \leq t$ $\forall a \in \sum$: $\min\{|x|_a,$
$k\}$ occurrences of a in x are replaced by some word in $h_i(a)$ to
obtain y from x;

Uniformly k-limited (ukIT0L) systems: $x \Rightarrow y$ iff $\exists 1 \leq i \leq t$:
$\min\{|x|,k\}$ symbols in x are replaced according to h_i to obtain y
from x.

Example 2:
Let the following concrete system be considered in more detail:
$G = (\{a, b\}, \{\{a \rightarrow aa, b \rightarrow ab\}\}, abb)$
    Derivation sequences when interpreting G as 0L system or as
some form of partial parallel grammar are as follows:
0L system $abb => aaabab => a^6 aba^2 ab$
0B system $abb => aabb => aaabab => a^3 abaab$

110L system $abb \Rightarrow aaabb \Rightarrow aaaabab \Rightarrow a^4ba^3b$
210L system $abb \Rightarrow aaabab \Rightarrow a^5abaab$
u210L system $abb \Rightarrow aabab \Rightarrow a^4bab \Rightarrow a^6bab$

In summary Parallel derivation grammars, also called L systems, can be classified in the following ways:

a.   Context-sensitive (IL) systems.
b.   Context-free (0L) systems.
c.   Deterministic (DL) systems.
d.   Propagative (PL) systems.
e.   Systems with extensions (EL systems).
f.   Systems with tables (TL systems).

4.1.4   Applications of L-System

In actual fact, Lindenmayer systems can be viewed, besides finite automata and context-free grammars, as the most useful "inventions" of the whole field of formal languages. Notably, they have been applied in computer graphics and developmental biology [19, 41,106].

-The derivation process of a certain grammar G working with an alphabet containing among other symbols, f, F, + and - is observed.

-A generatable string containing $/$, F, + and — is sequentially interpreted by a so-called "turtle" which is basically a pen equipped with a "direction"; the special symbols are interpreted as commands signifying:

Drawing: If a turtle reads an F, it draws a line of unit length by "walking" in the current direction.

Skipping: The turtle interprets f as: move the cursor (representing the turtle) forward by one unit without drawing.

Turning right: On seeing a +, the turtle turns right by δ degrees;

Turning left: analogously, a left turn by δ *degrees* is done when "seeing" input symbol -.

Often, the sequence of pictures drawn by a turtle (corresponding to the derivation sequence in question) is of interest, especially when "fractal properties" of the pictures concerned show up when interpreting sentential forms are continued. Then, an appropriate rescaling of the pictures in the sequence becomes an issue.

This sort of graphics is very popular: it is one of the basic features of the programming language LOGO designed to teach children the basics of programming, as described in [14]. In fact, the pictures that to be created using the interpreter for turtle graphics can be converted to any format.

4.1.5 Turtle graphics vs L-System

Seymour Papert invented "Turtle graphics" as a system for translating a sequence of symbols into the motions of an automaton (the "turtle") on a graphics display [126]; the original idea was to provide a programmable object that children could learn to think geometrically with). This turned out to be an ideal system for giving a geometrical interpretation to the dynamics of L-systems.

Turtle Graphics [50,126] is a method of drawing in which turtle commands are interpreted as drawing instructions. The turtle can be thought of as a moving pen. The LOGO programming language is probably the most familiar turtle graphics system.

The turtle state consists of a position and a heading. Given a step size and an angle increment, the turtle can respond to commands. Here is a description of implementation of turtle graphics that has been used in the software for generating the fractals in this paper.:

Angle increment = 2 * π / number of directions (radians). If number of directions = 4, then angle increment = 90°.

F : Pen down, Move forward. Moves and draws a line.

f : Pen up, Move forward. Moves without drawing line.

| : Reverse turtle direction. Rotates head by 180°.

+ : Turn left. Rotates head counter-clockwise by angle increment.

- : Turn right. Rotates head clockwise by angle increment.

< : Decrease number of directions. If number of directions becomes zero, set number of directions to maximum number of directions.

> : Increase number of directions. If number of directions becomes greater than the maximum number of directions, set the number of directions to 1.

[: Push turtle state. Saves turtle state on the stack. The position and heading are saved.

] : Pop turtle state. Restores turtle state to the state saved on top of the stack. The position and heading are restored. If the stack is empty then Pop generates a turtle error. This implies that Push and Pop instructions must be balanced.

4.1.6 Generation of fractal figures

In this book, many fractal figures are generated using L-System approach and interpretation in a Turtle graphic software developed by the us.

L-System Code	Generated Figures

Koch1 {
  Angle 6
  Axiom F--F--F
  F=F+F--F+F }	

Dragon {
  Angle 8
  Axiom FX
  F=
  Y=+FX--FY+
  X=-FX++FY- }	

Circle{
  Dirs = 14
  Axiom = F+F+F
  #Iterations = 5
  F=
+F+F+F+F+F+F+F+F+F+F+F+F+
  F+F+F+F+F|+F+F+F+F+F+F+F+F+
F+F+F+F+F+F+F }	

Circle1{
  Dirs = 15
  Axiom = F
  #Iterations = 5
  F=
+F+F+F+F+F+F+F+F+F+F+F+F+F+F
+F+F }	

Circle{
  Dirs = 8
  Axiom =Ff+F+fF
  #Iterations = 5
  F =
+F+F+F+F+F+F+F+F+F+F+
  F+F+F+F+F+F+F+F+F+F+F+F+F+F
+F+F+F+F+F+F+F }	

Circle{
  Dirs = 15
  Axiom = F, F
  #Iterations = 5
  F=
+F+F+F+F+F+F+F|+F+F+F+F+F+F+
F}	

LetterM{
  Dirs = 5
  Axiom = F
  #Iterations = 3
  F = +F+++F|++F--F
};LetterM	

Plant07 {
  axiom Z
  Z=ZFX[+Z][-Z]
  X=X[-FFF][+FFF]FX
  angle 14 }
	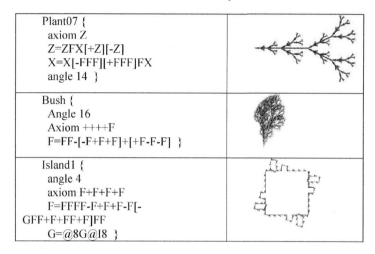

Bush {
  Angle 16
  Axiom ++++F
  F=FF-[-F+F+F]+[+F-F-F] }	

Island1 {
  angle 4
  axiom F+F+F+F
  F=FFFF-F+F+F-F[-
GFF+F+FF+F]FF
  G=@8G@I8 }	

Table 4.1: Fractal figures generated using L-System

Some of the figures have been shown in Table 4.1. Detailed L-system codes are given in Appendix-A for reference. A C program also has been developed to generate various kind of trees based on L-System.

4.1.7   About L-System

Lindenmayer Systems, popularly known as L-System in short, are string-rewriting systems that consist of an initiator (axiom), and a generator (rules).

The rules are applied to the axiom to generate a new string. This generation can be iterated to produce strings of arbitrary length. This string can be interpreted as a string of turtle graphics commands instructing a turtle graphics system to draw a picture.

If the axiom is F, and there is one rule F=F+F—F+F, then one iteration results in the string F+F—F+F. A second iteration results in the string F+F—F+F+F+F—F+F—F+F—F+F+F+F—F+F.

If F commands a turtle to move forward one unit, drawing a line, and + commands the turtle to turn left, and - commands the turtle to turn right, and the angle increment is 60°, then the axiom would produce a straight line.

4.1.8   An L-system example

KOCHACTUAL {
   Dirs = 6
   Axiom = F--F--F

#Iterations = 3
F=F+F--F+F
}; KOCHACTUAL

First four expansions of this L-System run by the software are shown in figure (Fig.4.1) below.

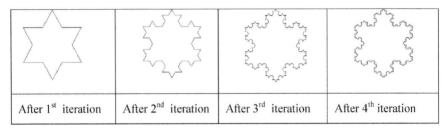

| After 1st iteration | After 2nd iteration | After 3rd iteration | After 4th iteration |

Figure 4.1: First four expansion of Koch curve

4.1.9   Representing mathematical sequence in L-System

This section illustrates the representation of mathematical series in L-System.

Example 1: To generate Fibonacci sequence
Consider the simple L-System grammar that is defined as follows ...
 Variables :   A , B
 Constants :   none
 Axiom   :   A
 Rules   :   A -> B
         B -> AB

This L-System produces the following sequence of strings ...
 Stage 0 : A
 Stage 1 : B
 Stage 2 : AB
 Stage 3 : BAB
 Stage 4 : ABBAB
 Stage 5 : BABABBAB
 Stage 6 : ABBABBABABBAB
 Stage 7 : BABABBABABBABBABABBAB

If we count the length of each string, we obtain the well-known Fibonacci sequence of numbers :
  1 1 2 3 5 8 13 21 34 ....

Example 2: To generate positive integers from 1 to n

The equivalent L-System may be as per the following:
Variables : A,B
Constants : any lowercase letters or digits,e.g., a,b,c... or 1,2,3.... etc.
Axiom : A
Rules : A => B, B=> Ba(where a is any constant)
This L-system produces the following sequence of strings ...
    Stage 0 : A
    Stage 1 : B
    Stage 2 : Ba
    Stage 3 : Baa
    Stage 4 : Baaa
    Stage 5 : Baaaa
    Stage 6 : Baaaaa
    Stage 7 : Baaaaaa   and so on.
If we count the length of strings generated from stage 1 onwards, we can get the positive numbers from 1 to infinity.

Example 3: To generate n numbers of positive odd integers.

The equivalent L-System may be represented as follows.
Variables : A,B
Constants : any lowercase letters or digits, e.g., a or 1,2,3.. etc
Axiom: A
Rules : A => B, B=> Baa(where a is any constant)

This L-system produces the following sequence of strings ...
    Stage 0 : A
    Stage 1 : B
    Stage 2 : Baa
    Stage 3 : Baaaa
    Stage 4 : Baaaaaa
    Stage 5 : Baaaaaaaa
    Stage 6 : Baaaaaaaaaa
    Stage 7 : Baaaaaaaaaaaa   and so on.
If we count the length of strings generated from stage 1 onwards we can get the positive odd numbers from 1 to infinity.

Example 4: To generate positive even integers.

The equivalent L-System grammar may be as per the following.

Variables : A,B

Constants : any lowercase letters or digits, e.g., a or 1,2,3.. etc

Start : A

Rules : A => Ba, B=> Baa(where a is any constant)

This L-system produces the following sequence of strings ...

    Stage 0 : A

    Stage 1 : Ba

    Stage 2 : Baaa

    Stage 3 : Baaaaa

    Stage 4 : Baaaaaaa

    Stage 5 : Baaaaaaaaa

    Stage 6 : Baaaaaaaaaaa

    Stage 7 : Baaaaaaaaaaaaa   and so on.

If we count the length of strings generated from stage 1 onwards, we can get the positive even numbers from 1 to infinity.

4.2   The Approach

4.2.1   Assumptions

As the symbol 'F' is considered as the only graphical symbol in turtle graphic, the L-System involving non-graphic symbols like A, B, C etc in its axiom as well as productions are not graphically interpreted. Koch curve represented in L-System using the symbol 'F' can easily be visualized by manual method or by any standard turtle graphic interpreted software for visualization. L-System for Koch curve and L-System for any mathematical sequence, e.g., Fibonacci sequence as discussed in the section 4.1.9, has been mentioned in the following table (Table 4.2) for ease of understanding. As Koch curve can be generated by using its L-System, graphical interpretation for mathematical sequence involving the symbols other than the symbol 'F' is simply omitted in Turtle graphics.

L-System for Fibonacci sequence	L-System for Koch curve
Variables : A , B	Variables : F
Constants : none	Constants : None
Axiom : A	Axiom : F--F--F
Rules : A -> B	Rules : F -> F+F--F+F
B -> AB or B-> A | B	

Table 4.2: L-System codes for Fibonacci sequence and Koch

Graphical interpretations of any L-System can be viewed in computer display after representing them in a particular turtle graphic interpretation. The visual representation of these L-Systems is shown in the table (Table 4.3) for reference after representing these L-Systems in Lsysedit turtle graphic source codes stored in separate files with file extension.l. It is observed that L-System using the graphic symbol 'F' generates Koch curve whereas the L-System using the non-graphic symbols 'A' and 'B' does not generate any figure.

L-System code for Fibonacci sequence	Figures generated
FIBONACCI{ Dirs=6 Axiom = A #Iterations=3 A=B B=AB };FIBONACCI	No Figure
L-System code for Koch curve	Figure Generated After 6th iteration
KOCHACTUAL{ Dirs = 6 Axiom = F--F--F #Iterations = 3 F=F+F--F+F };KOCHACTUAL	

Table 4.3: Visual representation of Koch curve

Keeping view of these facts as mentioned in the tables (Table 4.1 and Table 4.2) above, it is assumed that some interesting fractals may be generated by combining the L-System used for Koch curve with that of the L-System for mathematical sequence. As there is no such work found in literature how to combine any two L-Systems, a new way of combining these L-Systems is proposed to observe the graphical visualization. Following section presumes some hypothetical combination rules of any two L-Systems.

4.2.2   *Combination of L-Systems*

L-System used in section 4.1.8 for Koch curve is combined with L-System used in section 4.1.9 (Example 1), for Fibonacii sequence, to form a new set of axioms and rules. Assumptions regarding these hypothetical combinations are the following [8,125]:

Assumption I: One can combine two L-Systems say, L1 and L2, to form a new L-System, say L3, by associating the corresponding axioms and production rules designed for each L-System.

Assumption II: The axiom used in one L-System say L2, may be placed either in the left or in right of the symbol used in the axiom of the other L-System, say L1. In this chapter L1 represents the L-System for any mathematical sequence, L2 represents L-System for the classical fractal Koch curve and L2

Assumption III: As one cannot use two symbols in the left hand side of the production rule used in any L-System, due to the parallel grammar property of L-System, the graphic symbol 'F' cannot be placed in the left hand side of the new production rule. Only non-graphic symbol is used in the left hand side of the production rule.

Assumption IV: The axiom string of any L-System, L2, say Koch curve, may be placed or appended, in the left or right hand side of the axiom string of the other L-System, L1, say any mathematical sequence. If there is one axiom string and one production rule in the L-System L2, and there are two non-graphic symbols present in the L-System, L1, the axiom of L2 may be appended at the right or left hand side of the first non-graphic symbol present in the axiom of the L-System, L1, to form the new hybrid axiom. Similarly the production rule of the L-System, L2, may be appended at the right or left hand side of the second non-graphic symbol of the L-System, L2, to form the new hybrid production rule. The axiom string of L2 and the production string of L2 are to be appended in similar fashion with each occurrence of corresponding first and second non-graphic symbols used in the L-System, L1, except at the left hand side single symbol, which is used in the production rule of the L-System, L1.

Assumption V: The number of axioms present in the L-System, L1, should be same as that of the number of axioms present in the L-System L2.

Assumption VI: Number of non-graphic symbols should not exceed two for better visualization. One can use more than two non-graphic symbols in case of the L-System, L1, and does use more than one production rules. In general, if there are 'n' production rules to be required for the L-System, L2, then the L-System, L1, for mathematical sequence having n+1 number of non-graphical symbols may be appropriate for uniform association of axiom and production rules with non-graphic symbols.

Assumption VII: If L1 is an L-System and L2 is another L-System, the above-assumed combinations, as per the Assumption IV, generate a new L-System, L3.

Following section describes the application of these assumptions as discussed by taking L1 as the L-System for mathematical sequence and L2 as the L-System for Koch curve to generate possible L-System at a time as L3. The newly formulated L-System, L3, is named hybrid L-System in this book.

4.2.3   The new L-System or the Hybrid L-System

One can combine any L-System for classical fractals with that of the L-System for any mathematical sequence represented in L-System to formulate the new Hybrid L-System [8,125]. For example, let L1 represent the L-System for the mathematical sequence for generating Fibonacci numbers and L2 represent the L-System for the Koch curve, the possible new L-systems which may be formulated are mentioned in the table (Table 4.4).

With reference to the table (Table 4.2) few possible combinations of the two L-Systems are mentioned in the table (Table 4.4), although other combinations may be possible.

Categories	Associated Axiom	New Rules
1	AF—F—F	A=BF+F--F+F
		B=AF- -F--FBF+F--F+F
2.	F- -F--FA	A=F+F--F+FB
		B=F- -F--FAF+F--F+FB
3.	AF- -F- -F	A=B
	OR	B=AB F+F--F+F
	F- -F--FA	OR A=B B=A|B F+F--F+F
		OR A=B F+F--F+F
		B=AF—F—F | B F+F--F+F

Table 4.4: Combination of axiom and rules

These three categories are named as suffix-Koch category, prefix-Koch category and substitute-Koch categories for easier reference. Other non-growing combinations are not considered for category 3 of the above table. Next section elaborates an algorithm for formulating the new L-System for generation of different fractals.

4.2.4   The Algorithm

An algorithm [8,125] is suggested with some steps, not necessarily sequential, such as, the first to develop the recursive mathematical series using any number of variables, second to represent the same in L-System, the next to take any classical fractal's axiom and rules in L-Systems, and last to combine the axiom of the mathematical sequence with that of the fractals taken as base. After combining the axiom of standard fractals in L-System with that of the axioms and rules of L-System designed for representing mathematical sequences, the newly formed L-System may be tested with any turtle graphic L-System implemented software to generate this class of fractals. The proposed algorithm and all these steps are discussed with few examples below in the table (Table 4.5).

These three categories are named as suffix-Koch category, prefix-Koch category and substitute-Koch categories for easier reference. Other non-growing combinations are not considered for category 3 of the above table. These hybridized L-Systems have been tested with the software to produce various new types of fractals.

Step-1	Formulation of base fractal in L-System
Step-2	Formulation of discrete mathematical series represented in L-System
Step-3	Formulation of new hybrid L-System by combining the axiom and rules of fractals as generated in Step-1 and Step-2
Step-4	Repetition of Step-5 through Step-6 for Dirs = 3 to Maximum directions with step 1
Step-5	Running the new L-System
Step-6	Storing the image

Table 4.5: Proposed Algorithm written with permission from the paper titled "Growing a class of fractals based on the combination of classical fractal and recursive mathematical series in L-system" published by Machine Graphics & Vision

4.3 Experimentally Generated Fractals

4.3.1 Fractal figures for Fibonacci sequence and Koch curve

Assuming the first L-System for Fibonacci sequence as L1 and the second L-System for Koch curve as L2, and applying the proposed combination rules as suggested in the table (Table 4.3), the Hybrid L-System as L3 is formulated. This new Hybrid L-System is named as FibonacciKoch for reference. The axiom, the production rules and the formulated L-System are mentioned in the table (Table 4.6).

L-System	L1: Fibonacci Sequence	L2: Koch Curve	L3: FibonacciKoch atright	L-System code
Axiom	A	F- - F- - F	AF - - F--F	FibonacciKochSuffix { Dirs = 6
Production Rules	A = B B = AB	F =F+F - - F + F	A=B F+F--F+F B=AF - - F- -FB F+F--F+F	Axiom = AF--F--F A=BF+F- - F+F B=AF--F--FBF+F--F+F}; FibonacciKochSuffix

Table 4.6:Formulated L-System codes

In the next table (Table 4.7) each column heading represents the number of iterations, while each row headings represents the number of directions in which the turtlehead is to be rotated.

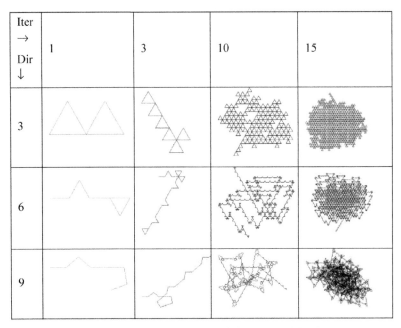

Iter → Dir ↓	1	3	10	15
3				
6				
9				

Table 4.7: Various figures generated using combinations

4.3.2   *Fractal figures for Mathematical series 1 to n and Koch curve*

Assuming the first L-System for Mathematical sequence of infinite series from 1 to n number sequence as L1 and the second L-System for Koch curve as L2, and applying the proposed combination rules as suggested in the table (Table 4.3) one can formulate the Hybrid L-System as L3. This new Hybrid L-System is named as FibonacciKoch for reference. The axiom, the production rules and the formulated L-System are mentioned in the table (Table 4.8).

L-System	L1: OnetoNsequence	L2: Koch Curve	L3: OnetoNKochatright	L-System code
Axiom	A	F- - F- - F	AF - - F--F	OnetonKochSuffix{ Dirs = 6
Constant	B	F =F+F - - F + F	A= AF--F- -FBF+F-- F+F	Axiom = AF--F--F
Production Rules	A = AB			A=AF--F--FBF+F--F+F }:OnetonKochSuffix

Table 4.8: L-System codes for 1 to infinity series

Iter → / Dir ↓	1	3	4	Converged Figure
3				After 3 iterations
6				After 3 iterations
9				After 6 iterations
12				After 3 iterations
15				After 15 iterations

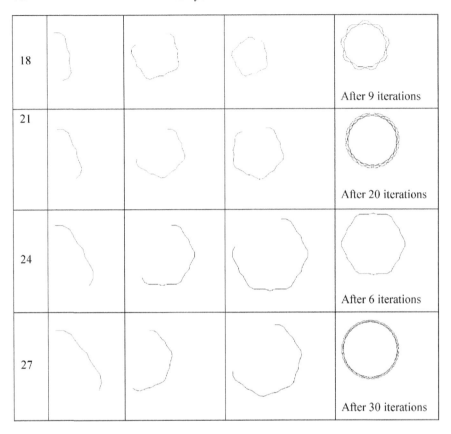

18				
				After 9 iterations
21				
				After 20 iterations
24				
				After 6 iterations
27				
				After 30 iterations

Table 4.9: Converged figures reproduced with permission from the paper titled "Growing a class of fractals based on the combination of classical fractal and recursive mathematical series in L-system" published by Machine Graphics & Vision

As there are an infinite number of fractal figures generated like the Fibonacci sequence, interesting converged figures are shown above in the table (Table 4.9) for reference.

4.3.3   Fractal figures based on different combinations

As many of such types of fractal figures are generated, a few fractal figures using the possible proposed combinations with Fibonacci sequence series have been tested and the figures as generated are shown in the table (Table 4.10).

L-System Code	Iterations = 1	Iterations = 3	Iterations = 5	Iterations = 10

suffixKoch{
  Dirs = 6
  Axiom = AF--F--F
  A=BF+F--F+F
  B=AF--F--FBF+F--F+F
}:suffixKoch				

prefixKoch{
  Dirs = 6
  Axiom = F--F--FA
  A=F+F--F+FB
  B=F--F--FAF+F--F+FB }:prefixKoch				

SubstiKoch{
  Dirs = 6
  Axiom = AF--F--F
  A=B
  B=ABF+F--F+F
}:SubstiKoch				

Table 4.10: Fractals generated for new combination of axiom and rules

4.4   Variations of Koch Curves

Alternate Koch curve L-systems with variations of directions as well as prefix, suffix and substitute approach to recursive mathematical formulae can generate fractals also. Some possible alternate Koch curves are also tested to grow fractals. A list of Koch alternatives is mentioned in the table (Table 4.11) for reference. As first example, given in section 4.1.8, is named as Koch, other alternatives are named as Koch1, Koch2, Koch3 and Koch4 respectively.

L-System variations	Iterations = 1	Iterations = 4	Iterations = 6
Koch1{ Dirs = 6 Axiom = F-F-F-F F = F-F+FF-F-F+F };Koch1			
Koch2{ Dirs = 12 Axiom = F---F---F---F F = -F+++F---F+ };Koch2			
Koch3{ Dirs = 4 Axiom = F-F-F-F F = F-F+F+FF-F-F+F };Koch3			
Koch4{ Dirs = 4 Axiom=F+F+F+F F=F-FF+FF+F+F-F-FF+F+F-F-FF-FF+F};Koch4			

Table 4.11 : Fractals generated after some iterations of L-System implemented alternate Koch curves

Any additional user designed alternatives may be used to grow fractals also. A few generations of fractals are shown in Table 4.12, taking these alternatives of the Koch curve that is mentioned in the first column of Table 4.11, a class of fractals are also generated combining Koch1 (see first column of Table 4.10) and Fibonacci sequence L-System as per Table 4.3, are shown in table (Table 4.12). Detailed codes are available in the Appendix-B as per names mentioned in the first column of the table (Table 4.12).

L-System Code	Iterations =1	Iteratio ns=2	Iterations = 3	Iterations=1 0	Iterations=18
Koch1Suffi xFibonacci					
Koch1Prefi xFibonacci					
Koch1Subst iFibonacci					

Table-4.12: Fractals combining Koch1 and Fibonacci sequence L-System

4.5 Fractals with other Mathematical Sequences

Combining the axiom of Koch curve or any alternatives on Koch, such as Koch1, Koch2, Koch3, Koch4 etc., with the L-System of mathematical sequence at Example-2, Section 4.1.9 (i.e., sequence of positive integers from 1 to n, named as Oneton in the first column of Table 4.12), generate many interesting fractals. Out of the possible combinations some fractals are shown in Table 4.12. In all these mathematical sequences, from Example-2 to Example-4 at Section 4.1.9, the letter "X" throughout the generation process represents a constant value. Associating Koch axiom with that of mathematical sequence for generating positive integers from one to infinity many new fractals are generated. For easier reference, L-System for these combinations are named as KochSuffixOneton, KochPrefixOneton etc., in a format of XYZ, where X may take the base fractal's axiom, Y may take the possible category and Z may take the mathematical series name.

L-System Name	Iterations =1	Iterations =2	Iterations=5	Iterations=10
KochSu ffixOne ton				
KochPr efixOne ton				
KochSu bstiOne ton				
Koch1S uffixOn eton				
Koch1P refixOn eton				
Koch1S ubstiOn eton				

Table 4.13 : Fractals using Koch and Koch1 and recursive series one to n

Other mathematical sequences developed in examples (Example-3 and Example-4 of Section 4.1.9) generate the same fractals as in Table 4.13. Since the value of X of L-System code for mathematical sequence in those examples is assigned a blank value, the generated turtle graphic leaves no visible trail. So, a combination of any given fractal's axiom and rules (of any alternatives), with all those mathematical sequences as described in this book will definitely generate the same fractals.

L-System Name	Iterations =1	Iterations =2	Iterations=5	Iterations =10
SierpinskiSuffixOneton With X=F-F-F-F-F				
NilSubstiEven	No figure			
PeanoSuffix Fibonacci				
PeanoPrefixF ibonacci				
BushSuffixFi bonacci				
BushPrefixFi bonacci				

Table 4.14: Fractals combining other base fractals and same mathematical sequences

A few fractals are shown in Table 4.14 by applying our proposed approach. Some base fractals other than Koch curves with the one to *n* series have been shown in Table 4.14. The proposed approach can generate an infinite number of these hybrid fractals [8,125]. This is reproduced with permission from the paper titled "Growing a class of fractals based on the combination of classical fractal and recursive mathematical series in L-system" published by Machine Graphics & Vision

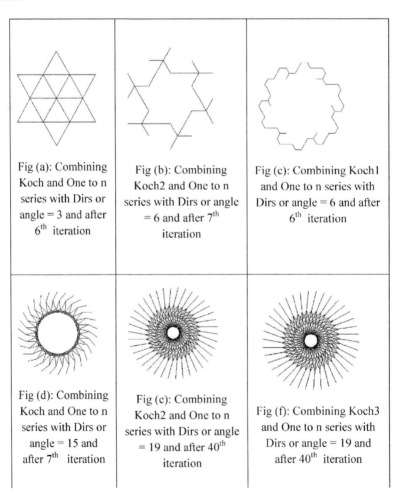

Fig (a): Combining Koch and One to n series with Dirs or angle = 3 and after 6th iteration	Fig (b): Combining Koch2 and One to n series with Dirs or angle = 6 and after 7th iteration	Fig (c): Combining Koch1 and One to n series with Dirs or angle = 6 and after 6th iteration
Fig (d): Combining Koch and One to n series with Dirs or angle = 15 and after 7th iteration	Fig (e): Combining Koch2 and One to n series with Dirs or angle = 19 and after 40th iteration	Fig (f): Combining Koch3 and One to n series with Dirs or angle = 19 and after 40th iteration

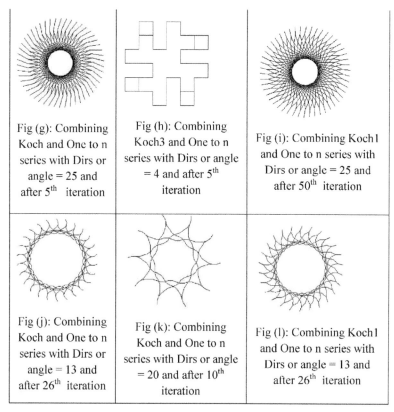

Fig (g): Combining Koch and One to n series with Dirs or angle = 25 and after 5th iteration	Fig (h): Combining Koch3 and One to n series with Dirs or angle = 4 and after 5th iteration	Fig (i): Combining Koch1 and One to n series with Dirs or angle = 25 and after 50th iteration
Fig (j): Combining Koch and One to n series with Dirs or angle = 13 and after 26th iteration	Fig (k): Combining Koch and One to n series with Dirs or angle = 20 and after 10th iteration	Fig (l): Combining Koch1 and One to n series with Dirs or angle = 13 and after 26th iteration

Table 4.15 : Converged fractal figures combining Koch variations and Mathematical series for infinite positive integers

A few novel fractals are generated using combinations of Koch curve and its variation with that of 1 to n series. As these types of figures are totally new to the world of fractals, these figures may be added to the future gallery of fractals. These figures are shown in the table (Table 4.15). The detailed codes in L-Systems are given in Appendix-B.

4.6   Interpretation of Result

4.6.1   Comparison of Koch curve with Hybrid system

As graphical visualization for Fibonacci sequence is not possible in turtle graphics due to the absence of graphic symbol 'F' in the axiom as well as production rule the comparative study between L-System of Koch curve and L-system of Hybrid L-System as shown in the table (Table 4.6) are mentioned in the table (Table 4.16) for reference. For better understanding the L-system of Koch curve and L-System code after combination have been given below.

The L-System code for Koch curve is

Axiom : F- - F- - F

Rule of Production :

Dirs = 6

F = F+F--F+F

The L-System for the Hybrid curve after combination of Koch and Fibonacci is

Dirs = 6

Axiom = AF--F--F

Productions:

A=BF+F--F+F

B =AF--F--FBF+F--F+F

The above two L-Systems generate different strings after each iteration. In the case of Koch curve the symbol 'F' is expanded every time where as in Hybrid system the non-graphic symbol helps in growing the production string. The growth of L-system curve for Koch curve has been shown in the Table 4.17 where as the hybrid L-system for combination of Koch and Fibonacci sequence has been shown in the Table 4.18. In both of these tables (Table 4.17 and Table 4.18) the symbol 'F' in the axiom is substituted by the production string with same usual rate of reduction 1/ 3 every time.

Fractal	L-System	Figure After 1st iterations	Figure after 2nd iterations	Convergent Figure
Koch curve Reference Table 4.6	Koch{ Dirs = 6 Axiom = F--F--F F=F+F--F+F};			
Fibonacci Koch Reference :Table 4.6	FibonacciKochSuffix { Dirs = 6 Axiom=AF--F--F A=BF+F- -F+F B=AF--F--FBF+F-- F+F};			

Table 4.16 : Comparison between Koch and Hybrid L-System figures

String after 1st Iterations	String after 2nd iterations	String after 3rd iterations
F + F- - F + F - - F + F- - F + F - - F + F- - F + F	F+F--F+F + F+F-- F+F - - F+F--F+F + F+F--F+F - - F+F-- F+F + F+F--F+F - - F+F--F+F + F+F-- F+F - - F+F--F+F + F+F--F+F - - F+F-- F+F + F+F--F+F	F+F - - F + F+F+F - - F + F - -F+F - - F + F+F+F - - F + F + F+F - - F + F+F+F - - F + F- - F+F - - F + F+F+F - - F + F - - F+F - - F + F+F+F - - F + F—F+F - - F + F+F+F - - F + F + F+F - - F + F+F+F - - F + F- - F+F - - F + F+F+F - - F + F - - F+F - - F + F+F+F - - F + F+F+F - - F + F—F+F - - F + F+F+F - - F + F + F+F - - F + F+F+F - - F + F- - F + F--F+F - - F + F+F+F - - F + F - - F+F - - F + F+F+F - - F + F--F+F - - F + F+F+F - - F + F + F+F - - F + F + F--F+F - - F + F + F+F+F - - F + F + F+F - - F + F F+F - - F + F+F+F - - F + F--F+F - - F + F+F+F - - F + F - - F+F - - F + F+F+F - - F + F--F+F - - F + F+F+F - - F + F + F+F - - F + F+F+F - - F + F + F--F+F - - F + F+F+F - - F + F
Graphical Interpretation	Graphical Interpretation	Graphical Interpretation

Table 4.17 : String Growth of Koch curve

In case of hybrid L-system as shown in the Table 4.18, the value of 'A' in production is substituted in the axiom. Subsequent growth of strings has been shown in the Table 4.18. The equivalent graphical interpretations are also given.

String after 1st Iterations	String after 2nd iterations	String after 3rd iterations
BF+F- -F+F F- -F- -F	(AF- -F- - F B F+F - - F +F)F+F- -F+F F- -F- -F	(BF+F- - F+F) F- -F- - F (AF - -F- -FBF+F- -F+F) F+F - - F +F)F+F- -F+F F- -F- -F
Graphical Interpretation	Graphical Interpretation	Graphical Interpretation

Table 4.18 : String Growth of Hybrid curve

4.6.2 Arbitrary Figures

L-System	L1: Fibonacci Sequence	L2: Koch Curve	L3: FibonacciKoch atright	L-System code
Axiom	A	F- - F- - F	AF - - F—F	FibonacciKochSuffix{ Dirs = 6
Production Rules	A = B B = AB	F =F+F - - F + F	A=BF- - F- - F B=AF - - F- - FB F+F--F+F	Axiom = AF--F--F A=BF - - F- - F B=AF--F--FBF+F--F+F }:FibonacciKochSuffix

Table 4.19: Arbitrary substitution of strings in Hybrid L-System

Iter → Dir ↓	1	4	10	15
3				
6				
9				
12				
15				
18				
21				

Table 4.20: Arbitrary figures after arbitrary substitution of axiom strings

For arbitrary string substitution one can generate some fractal figures as shown in the table 4.19. The above fractal figures do not follow the proposed approach as discussed in this chapter. If the axiom string of Koch curve is arbitrarily substituted at the right side of the non-graphic symbol in Fibonacci sequence, i.e., B, some fractal figures are generated. However one may not get a regular converged figure. Few arbitrary figures are shown in the Table 4.20 in order to establish the fact of irregularity.

4.7   Summary

L-systems can be used to generate fractals by applying a turtle interpretation to the symbols of the grammar. These same grammars can also be applied to generate many standard discrete mathematical series. The purpose of this chapter is to combine these two applications of L-systems to generate novel fractals by applying a turtle interpretation to the grammars that generate interesting mathematical series to build new fractals.

The approach suggested in this chapter for generating hybrid fractals could form the basis for modeling various types of growth phenomena found in nature. The relationship of the fractal dimension of the generated hybrid fractal to that of the base fractal needs to be explored. In subsequent growing process of these fractals are directly proportional to the strings associated with non-graphic symbols. It is observed that out of all mathematical sequences considered, Fibonacci sequence would always yield most space-filling fractals for the same number of iterations.

Chapter 5

L-System Strings from Ramification Matrix

The aim of this chapter is to provide a theoretical basis for designing a tree by calculating the equivalent ramification matrix. This chapter describes the methods to compile the equivalent ramification matrix to model a realistic tree. As L-System has become a more powerful and widely used tool for the creation of botanically accurate plants, an algorithm has been suggested to construct the tree from ramification matrix. The critical issue, i.e., how to derive equivalent L-system production rules from the ramification matrix has been discussed here. A theoretical background regarding the construction of ramification matrix from scratch is discussed.

The modeling of a tree is traditionally complex. The aim of this chapter is to provide a basis for designing a tree by calculating the ramification matrix. Naturally, tree modeling also has an application in botany. Using L-Systems, botanists are able to simulate the growth and development of plants, and are subsequently better able to understand the processes behind some of the complex tree structures found in nature. It also generally imposes restrictions on the practicality of modeling, as realistic looking trees are naturally complex and difficult to describe in an intuitive manner. This is shown clearly in L-Systems where the modeler requires substantial experience in order to model a realistic tree.

5.1   Definition of Terms

It is important to clarify some of the terminology used in L-Systems so as to avoid misunderstanding.

5.1.1   Modules

A module, in the context of L-Systems, is defined as "any discrete constructional unit that is repeated as the plant develops". In this book, a module is considered to be the encapsulation of a symbol, representing the name of the module, and a set of numerical attributes. The purpose of these attributes is discussed below. A string of modules representing a plant is referred to as an L-System string.

5.1.2   *Productions*

A production specifies the way in which modules in an L-System string are replaced. In a context-free system, a production consists of a predecessor defined by a single module, and a successor defined by a set of zero or more modules.

5.2   Parallel Rewriting Systems

At the heart of L-Systems is the idea of parallel rewriting systems. An L-System is defined by a starting string and a list of productions. Each module in the L-System string is compared with the predecessor module of each production. If a match is found, the module is replaced with the successor modules defined in the production. An example of this is given below.

$$\omega \ : \ BAB$$
$$p_1 \ : \ A \rightarrow C$$
$$p_2 \ : \ B \rightarrow D$$

In the above L-System, ω is the starting string, p_1 and p_2 are the productions in which A and B are the predecessors and C and D are the successors. In the first iteration, the module named A in the starting string would be replaced with the module named C, and the module named B would be replaced with the module named D. The resulting L-System string would therefore be DCD. This type of rewriting process is shown graphically in Figure 5.1.

Figure 5.1: A graphical rewriting of an L-System string

5.3   An Elementary L-System Parser

The implementation of this system is a relatively simple task. In this section we will discuss the structure of a module, followed by a look at an effective method of storing L-System strings, and finally the processes involved in writing an L-System parser.

5.3.1   The structure of an L-System module

An L-System module (a module, in the context of L-Systems, is defined as "any discrete constructional unit that is repeated as the plant develops), consists of a symbol and a set of attributes. This is conveniently represented as an object with the following structure(Table 5.1):

Attr. Name	Type	Description
Symbol	Char	A single character representing the symbol
numAttr[0]- numAttr[4]	double	An array of numerical attributes. The module may therefore contain at most five attributes.

Table 5.1: Module structure

5.3.2   L-System strings

An L-System string is a string of module objects. A useful structure for storing such a string is a Vector. A Vector is essentially an array with dynamic size. As elements are added to a Vector, so the Vector increases its maximum size and as elements are removed from a Vector, so the Vector frees the memory used by that element.

5.3.3   Rewriting the L-System string

A dual string may be implemented for the rewriting system whereby the rewritten string is stored in another Vector. In order to rewrite an L-System string all that needs to be done is to iterate through the provided L-System Vector, at each module comparing the symbol field with a set of predecessor symbols. If a match is found, a set of successor modules is placed into the new Vector. Once this process is complete, the provided L-System Vector is replaced with the new Vector. This is shown in the pseudo-code below:

```
while (there are modules in the provided L-System Vector)
module = lSystemVector.nextElement ()
switch (module.symbol)
        'F':newVector.addElement ('[')
                newVector.addElement ('F', 1.0)
                newVector.addElement (']')
        'A':newVector.addElement('A',module.param[0]-0.5)
lSystemVector = newVector
```

5.4   Generating The Tree

The tree structure is generated in a recursive manner as follows:

- Generate geometry for a branch
- Choose a branch order from the ramification matrix
- Rotate through a specified angle (twist and branching angle)
- Generate a tree of depth level specified by ramification matrix (introduced by Viennot *et al*, this technique provides a purely statistical approach to modeling binary trees
- Rotate through a specified angle (twist and branching angle)
- Generate a tree of depth level specified by ramification matrix

5.4.1   Parsing the L-System

Parsing the L-System to generate the plant geometry is accomplished in a similar manner to the rewriting process described above [9]. The symbol of each module is compared to a list of symbol, and if a match is found the corresponding set of instructions is executed. The fragment of pseudo-code below demonstrates this process:

```
While (there are modules in the provided L-System Vector)
module = lSystemVector.nextElement ()
switch (module.symbol)
'F': Draw a branch of length module.param[0]
'-': Rotate by module.param[0] degrees clockwise
'[': Push turtle state
']': Pop turtle state
```

Due to its rewriting nature, the system must be "re-grown" every time a change to a production is made. This can make modeling L-Systems a tedious task, and the design of a real-time, interactive L-System modeler nearly impossible.

5.5 Advantages and Disadvantages of L-Systems

The primary advantage of L-Systems is their ability to model complex plant structures through the definition of a few key productions. These productions, however, often take many hours of trial and error to produce, even by experienced L-System modelers. The rewriting nature of the system implements the "re-grown" formulae every time a change to a production is made. Therefore, the design of a real-time, interactive L-System modeler becomes a tedious task, and nearly impossible. In general, the fractal nature of L-Systems causes the complexity of the generated string to increase exponentially with every rewriting-iteration. After only a few iterations, the generated string can run to many pages, and the geometry produced can bring even a powerful system.

5.6 Ramification Matrices

Introduced by Viennot et. al., this technique provides a purely statistical approach to modeling binary trees. This section begins with a look at how ramification matrices are compiled.

5.6.1 Compiling a matrix

The following description follows closely that of Viennot, and demonstrates the relationship between the ramification matrix and the plant structure. Three types of ramification matrices can be generated. One for compiling the matrix from basic branching scheme, second for purely binary tree and the last one for non-binary tree. The following sections describe all these three types with examples.

5.6.2 Ramification matrix for basic branching scheme

Plants seem to exhibit arbitrary complexity while possessing a constrained branching structure. The topology of a plant is characterized by a recursive branching structure. Rick Parent in his Computer Animation Book proposed a two dimensional branching structure of interest are shown in Figure 5.2.

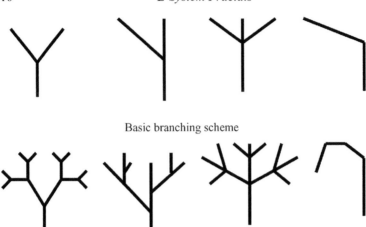

Basic branching scheme

Structures resulting from repeated application of a single branching scheme

Figure 5.2: Branching structures of interest in two dimensions

A sample structure from Figure 5.2 has been taken initially. This is 1ˢᵗ structure from the second row of the figure 5.2, which is an ordered one. By using the above structure, one can construct the following sample structure as shown in Figure 5.3, where nodes and branches are represented using numbers.

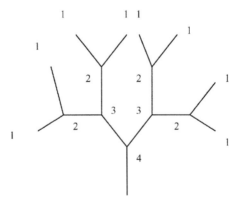

Figure 5.3: Ordered tree from basic branching schemes.

The following arbitrary tree as shown in Figure 5.4 closely follows that of Viennot and demonstrates the relationship between the ramification matrix and the plant structure.

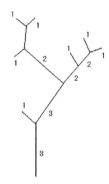

Figure 5.4: An arbitrary tree structure with ordered branches

The edges (branches) of the tree are labeled in the following recursive manner:

The terminal branches are labeled 1

If a branch has two children labeled i and j then the label of the branch is the larger of i and j if $i \neq j$

$i+1$ if $i = j$

The label of a branch is called the order of the branch, and is proportional to the size of the tree contained within its children. In the calculation of the actual values of the ramification matrix, we define the following quantities:

for $k \geq 2$, a_k is the number of order k branches

for $1 \leq i < k$, $b_{k,i}$ is the number of branches whose children have order k and i

for $k \geq 2$, $b_{k,k}$ is the number of branches whose children have order k-1 and k-1

for $1 \leq j \leq k$, $p_{k,j}$ is the probability of a branch has children of order k and j and is equal to $\dfrac{b_{k,j}}{a_k}$

Thus, the ramification matrix corresponding to Figure 5.4 with k rows and j columns would be:

$$\begin{bmatrix} 1 & & \\ 2/4 & 2/4 & \\ 1/2 & 0 & 1/2 \end{bmatrix}$$

Table 5.2: Ramification Matrix

where $p_{1,1}$ has just been included for the sake of completing the form of the matrix. As can be seen from this matrix, the probability of a branch of order 2 (k=2) having children of order 2 and 1 (j =1) is 50%, while the probability of having children of order 1 and 1 (j=2) is 50%.

Similarly the ramification matrix for Figure 5.3 is as given below:

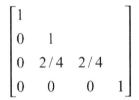

$$\begin{bmatrix} 1 & & & \\ 0 & 1 & & \\ 0 & 2/4 & 2/4 & \\ 0 & 0 & 0 & 1 \end{bmatrix}$$

Table 5.3: Ramification matrix

5.6.3   Generation of L-system string from the ramification matrix

For example in the above figure (Figure 5.3) the equivalent L-system re-writing rules to be generated. After applying the following algorithm one can generate the L-system grammar.

Algorithm

Step1: Rename the terminal edges and give any alphabetical name, i.e, A,B,C....

Step2 : Name the higher order edges with next higher alphabets, i.e., B,C...

Step 3: Repeat Step-2 for subsequent higher order edges until there is no more ordered branch to be renamed

Step 4: Note down the total strings generated from the root edge to the terminal edge.

Step 5: Construct the equivalent binary tree.

Step 6: Traverse the binary tree in preorder to get the final string

Step 7: Root node Alphabet may be the initial production and next branch will generate the set of production rules

Example:

As per the above algorithm we may generate the L-system grammar for the above tree with ramification matrix value as given in Table 5.2: Rename all terminal edges having order 1 of the tree with the letter B, order 2 with A, order 3 with C and order 4 as D. The complete string will be D = BAB C BAB BAB C BAB   for symmetric tree. As the common string is BAB we can replace the alphabet C with A. So the new string may be BAB A BAB A BAB A BAB. The equivalent binary tree may be constructed like the following.

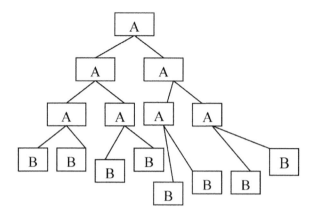

Figure 5.5: Equivalent binary tree with labels for the

ramification matrix given in Table 5.3

After applying the algorithm we can generate the L-system rewriting as ω = BAB and the production rules are

I.   B => BAB and
II. A => A.

From the ramification matrix tree can be generated.

5.6.4   Normalisation of the matrix

Because the values within the matrix represent branching probabilities, the sum of each row must be equal to 1. The function introduced above provides no guarantee that this will be the case, so we must normalise each of the values in the matrix.

This is fairly simply accomplished by adding each row and dividing each element in the row by the value of the sum. This maintains the ratios between the elements while ensuring that the sum of each row is exactly equal to 1.

5.7   Fractal Figures Using The Grammar

Novel fractals [7] can be generated using the grammar axiom and production rules using the concept of hybrid fractal generation algorithm as discussed in chapter 4. A set of fractal figures as discussed in chapter 4, is generated using Koch and its variation with that generated grammar.

For convenience of reading, the axiom with the nongraphic symbol A has been taken as the axiom, where the letter B has been used in the production. The grammar uses the axiom A and the production rule A ⇒ ABA. Some converged fractal figures as generated are shown in Table 5.4. This grammar may be used to model a simple binary tree. An infinite number of figures can be generated. Detailed L-system codes of the figures of Table 5.4 are given in Appendix-C for reference.

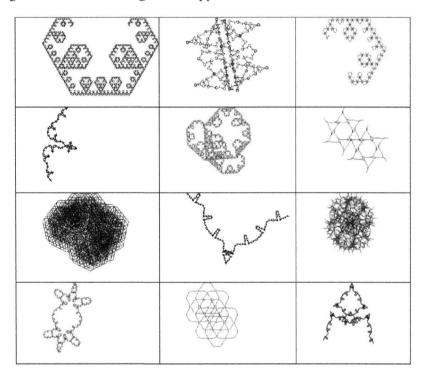

Table 5.4: Novel fractals using Koch curve and the grammar

5.8   Summary

Ramification matrices, due to their statistical nature, are able to store a tree-type rather than an actual tree model. A single matrix, consisting of only a handful of numbers, can describe a forest of similar trees, few of which are identical.

A major drawback to this method is the generation of the matrix itself. A ramification matrix can be compiled from an existing tree structure and while this can be useful in replicating real tree structures, few modelers have the inclination to count the branches in a tree and calculate branching probabilities. Quite a number of generic, pre-calculated, ramification matrices exist (such as binary and increasing-binary methods described by Viennot), but the fact remains that it is difficult to construct a ramification matrix from nothing.

Another drawback is the inability of ramification matrices to produce non-binary trees. At every branch point there is a split into exactly two branches. While this is suitable for the modeling of trees, it proves unsuitable for the modeling of small plants or flowers, as they often have more than one branch per node.

But the main objective of this ramification matrix and modeling of binary tree is to generate the grammar to produce hybrid fractals. In future, binary trees as well as non-binary trees can be modeled using the concept of ramification matrix.

Chapter 6

3D Modeling of Realistic Plants

This chapter introduces some key aspects of a plant modeling system for use in a three-dimensional real-time environment like practicality of modeling, realism of models produced and the rendering speeds obtained by these models. Various systems are implemented to demonstrate these key aspects in this chapter.

The modeling of plants is traditionally complex and the high number of polygons produced often makes them unsuitable for real-time environments. The goal of this chapter is to discuss the design and implementation of methods proposed by various researchers.

Plants are so common in human life that often nobody takes careful notice of them. It is only botanists and plant lovers who discuss the growth of these plants. The omni presence of plants makes us realize just how naked such an environment would appear in their absence. It is therefore very essential to have realistic plants in a virtual environment.

Almost since their inception, vehicle simulations and computer games have used plants extensively to improve the realism of their scenes. As computers and their graphical capabilities have become better and faster, so the plant models in such applications have become more detailed and realistic from single-polygon texture-mapped billboards to intricate meshes containing hundreds, even thousands of polygons.

Naturally, plant modeling also has an application in botany. Using L-Systems botanists are able to simulate the growth and development of plants and are subsequently better able to understand the processes behind some of the complex plant structures found in nature. Detailed aspects of plant development such as the amount of light received and the concentration of nutrients in the soil can be modeled in great detail. Even more abstract phenomena such as topiary (the art of pruning bushes and trees into shapes) can be accurately and easily modeled.

The modeling and rendering of plants is seldom a trivial task. In the design of a plant modeling system, several choices must necessarily be made as to whether the system focuses on the practicality of modeling, the realism of the resultant model or the speed at which it can be rendered. Rarely, a system is able to excel in all three criteria. A high degree of realism, by nature, increases the complexity of the model, thus reducing the speed at which is can be rendered. It also generally imposes restrictions on the practicality of modeling, as realistic looking plants are naturally complex and difficult to describe in an intuitive manner. This is shown clearly in L-Systems where the modeler requires substantial experience in order to model realistic plants.

This chapter begins with a discussion of the basics of plant modeling system. This is followed by the key factors required for design and implementation of a plant modeling system. Various methodologies proposed by researchers from time to time for the practical modeling of plants are discussed.

6.1   Related Work

The modeling and rendering of plants is not a new field of research and has received considerable attention in recent years. Some of the topics related to this, are mentioned in this section.

6.1.1   D0L-systems

The deterministic context-free class of L-systems is called D0L-systems. Formal definitions describing D0L-systems and their operation are as follows. More details [42,45,46] in this connection can also be seen.

Let V denote an alphabet, $V*$ the set of all words over V, and $V+$ the set of all nonempty words over V. A *string 0L-system* is an ordered triplet $G = <V, \omega, P>$, where V is the *alphabet* of the system, $\omega \in V^+$ is a nonempty word called the *axiom* and $P \subset V \times V*$ is a finite *set of productions*. A production $(a, \chi) \in P$ is written as $a \rightarrow \chi$. The letter a and the word χ are called the *predecessor* and the *successor* of this production, respectively. It is assumed that for any letter $a \in V$, there is at least one word $\chi \in V*$ such that $a \rightarrow \chi$. If no production is explicitly specified for a given predecessor $a \in V$, the *identity production* $a \rightarrow a$ is assumed to belong to the set of productions P. An 0L-system is *deterministic* (noted *D0L-system*) if and only if for each $a \in V$ there is exactly one $\chi \in V*$ such that $a \rightarrow \chi$.

Let us consider the earlier example as given in section 2.3.1. This can be represented by the set of rules as shown below.

$$\omega \ : \ b$$
$$p_1 \ : \ a \rightarrow ab$$
$$p_2 \ : \ b \rightarrow a$$

In the 1st derivation step i.e. the 1st step of rewriting, the axiom b is replaced by a using production $b \rightarrow a$. In the second step a is replaced by ab using production $a \rightarrow ab$. The word ab consists of two letters, both of which are *simultaneously* replaced in the next derivation step. Thus, a is replaced by ab, b is replaced by a, and the string aba results. In a similar way, the string *aba* yields *abaab* which in turn yields *abaababa*, then *abaababaabaab*, and so on

6.1.2   Edge Rewriting vs. Node Rewriting

Two modes of operation for L-systems with turtle interpretation, called *edge rewriting* and *node rewriting* are used for generating fractals. These terminologies are borrowed from graph grammars [1, 2]. In the case of edge rewriting, productions substitute figures for polygon edges, while in node rewriting, productions operate on polygon vertices. Both approaches rely on capturing the recursive structure of figures and relating it to a tiling of a plane. Although the concepts are illustrated using abstract curves, they apply to branching structures found in plants as well.

Edge rewriting can be applied to the *dragon curve* [22], where the L-system that generated the dragon curve can be as follows.

dragon curve{
    dirs = 4
    axiom = F_1
    $F_1 = F_1 + F_r +$
    $F_r = - F_1 - F_r$
}

Both the F_l and F_r symbols represent edges created by the turtle executing the "move forward" command. The productions substitute F_l or F_r edges by pairs of lines forming left or right turns. Many interesting curves can be obtained assuming two types of edges, left and right.

The idea of node rewriting is to substitute new polygons for nodes of the predecessor curve. In order to make this possible, turtle interpretation is extended by symbols, which represent arbitrary subfigures. During turtle interpretation of a string v, a symbol $A \in \tilde{A}$ incorporates the corresponding subfigure into a picture. To this end, A is translated and rotated to align its entry point and direction with the current position and orientation of the turtle. Considering the same example of dragon curve, the left figure L_{n+1} and right figure R_{n+1} are captured by the following formulae.

$L_{n+1} = L_n + R_n F+$
$R_{n+1} = -F L_n - R_n$

Suppose that curves L_0 and R_0 are given. One-Way of evaluating the string L_n (or R_n) for $n > 0$ is to generate successive strings recursively, in the order of decreasing value of index n. Thus, the generation of string L_n can be considered as a string-rewriting mechanism, where corresponding strings on the right side substitute the symbols on the left side of the recursive formulas. The substitution proceeds in a parallel way with F, + and - replacing themselves. Since all indices in any given string have the same value, they can be dropped, provided that a global count of derivation steps is kept. Consequently, string L_n can be obtained in a derivation of length n using the following L-system:

$\omega = L$
$L = L+ RF+$
$R = -FL -R$

Thus, the L-system for the dragon curve can be transformed into node-rewriting form as follows.

dragon curve{
   dirs = 4
   axiom = L
   L = L+RF+
   R = -FL-R
}

In practice, the choice between edge rewriting and node rewriting is often a matter of convenience. Neither approach offers an automatic, general method for constructing L-systems that capture given structures. However, the distinction between edge and node rewriting makes it easier to understand the intricacies of L-system operation, and in this sense assists in the modeling task. For example, some of the plant structures such as leaves may tend to fill a plane without overlapping [53,54]. Whereas a Bush can be a result of edge rewriting system, another tree can be generated out of a node rewriting system as may be seen in figure 6.1.

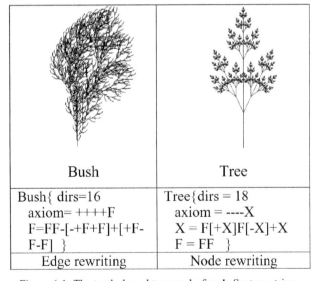

Bush	Tree

Bush{ dirs=16
  axiom= ++++F
  F=FF-[-+F+F]+[+F-
F-F] }
	Tree{dirs = 18
	  axiom = ----X
	  X = F[+X]F[-X]+X
	  F = FF }

Edge rewriting	Node rewriting

Figure 6.1: The turtle-based traversal of an L-System string

6.1.3   Stochastic L-Systems

All plants generated by the same deterministic L-system are identical. An attempt to combine them in the same picture would produce a striking, artificial regularity. Variation can be achieved by randomizing the turtle interpretation, the L-system, or both. Randomization of the interpretation alone has a limited effect. While the geometric aspects of a plant such as the stem lengths and branching angles are modified, the underlying topology remains unchanged. In contrast, stochastic application of productions may affect both the topology and the geometry of the plant. The following definition is similar to that of Yokomori [128] and Eichhorst and Savitch [98].

A *stochastic 0L-system* is an ordered quadruplet $G_{II} = <V, \omega, P, \pi>$. The alphabet V, the axiom ω and the set of productions P are defined as in an 0L-system. Function $\pi : P \rightarrow (0, 1]$, called the *probability distribution*, maps the set of productions into the set of *production probabilities*. It is assumed that for any letter $a \in V$, the sum of probabilities of all productions with the predecessor a is equal to 1.

6.1.4   Context-sensitive L-system

Productions in 0L-systems are context-free i.e. those are not concerned with the context in which the predecessor appears in the *L-systems*. However, sometimes the production application may also depend on the predecessor's context. This effect is useful in simulating interactions between plant parts due to the flow of nutrients or hormones. Various context-sensitive extensions of L-systems have been proposed [14, 28] and studied thoroughly in the past.

The following sample 1L-system makes use of context to simulate signal propagation throughout a string of symbols.

ω : *baaaaaaaa*
$p1 : b < a \rightarrow b$
$p2 : b \rightarrow a$

The first few words generated by this L-system are given below.

baaaaaaaa
abaaaaaaa
aabaaaaaa
aaabaaaaa
aaaabaaaa

....

It is clearly understood that a is replaced by b, only when b is the strict predecessor of a and also of-course b is replaced by a simultaneously in the previous string.

6.1.5   *Branching structures*

Branching structures are important in the modeling of plants, and can be easily implemented through the creation of a stack that holds the turtle state (position and orientation). Two modules, "[" and "]", are then defined to access the stack. The interpretation of the "[" module is to push the turtle state onto a stack, while the "]" module is used to pop the turtle state from the stack. A sample branching L-System is shown below, and is graphically represented in Figure 6.2. The turtle position, at each stage in the parsing process, is shown as an open circle.

F(2)[+(45)F(1)][-(45)F(1)]

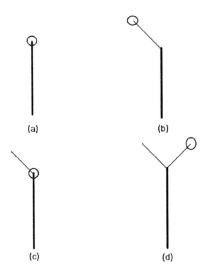

Figure 6.2: The turtle-based traversal of an L-System string

6.1.6   *Parametric L-Systems*

Parametric L-Systems make use of the numerical attributes of modules in the selection of an appropriate production. Productions are modified to include numerical conditions that must be satisfied by the module attributes in order for the production to be applied. This allows for the declaration of multiple productions with the same predecessor. These productions are distinguished only by differing numerical conditions. An example of a parametric L-System is given below.

ω : B(1)A(2,3)B(4)
p_1 : A(x, y) : y > 2 → C(1,2)
p_2 : A(x, y) : y <= 2 → B(5)A(x-1,0)
p_3 : B(x)   : x < 1 → B(x+1)
p_4 : B(x)   : x >= 1 → D(x-1)

The first module of the initial string, B(1), has a symbol, B, which can be matched to either p_3 or p_4. The numerical condition of p_3, however, requires that the module attribute be less than 1, which in this case is not so. The only production whose numerical condition is satisfied by the attribute of the module B(1) is p_4. The L-System string, after the first iteration, will therefore be: D(0)C(1,2)D(3). Once an L-System string has been generated, the corresponding geometry needs to be created. One popular method of accomplishing this, as described by Prusinkiewicz et. al., is to use a LOGO-style turtle. Each module in the L-System string is treated as a command to the turtle. Thus a module such as F(5) could be interpreted as "move the turtle forward by 5 units" and /(60) could mean "rotate the turtle by 60° in a counter-clockwise motion about the z-axis".

6.2 Key Factors

The requirements for the practical plant modeling system as suggested in the beginning can be put in the following three categories:

Practicality
The system must be easy to learn and intuitive to use

Realism
Plants generated must resemble real plants

Rendering speed
Plant models generated must be simple enough for use in a real-time environment where rendering speed is of great concern

6.2.1 Practicality

Plant modeling systems that require the user to have a vast knowledge of plant structures, and require the input of hundreds of parameters, are not usually trivial to master. A practical modeling system is a system that is simple and intuitive, and while easy-to-learn modeling systems often fail in the modeling of botanically accurate plant models, ease-of-modeling is often a greater priority. The practicality of a modeling system is often improved with the addition of a graphical user interface, allowing the user to model interactively.

6.2.2   Realism

The realism of models plays an important role in the development of a virtual environment, in order for the virtual environment to be believable. While plants are often represented by an abstract set of geometry (for example, a green sphere atop a brown cylinder) that is easy to model and rapid to render, "realism" refers to the creation of models that resemble actual plants.

6.2.3   Rendering speed

In the development of models for use in a real-time environment, the complexity of the models should be low enough to have acceptable rendering speeds. A standard practice can be made for real-time rendering by making frame rate (frames rendered per second) as ten or higher. A frame rate lower than this causes the animation to appear jagged to the user and the virtual environment may sacrifice much of its realism.

6.3   Component Modeling

Bernd Lintermann and Oliver Deussen introduced this modeling method that aims at separating the plant structure from the geometrical properties of the plant [95]. "Both tasks require their own optimized modeling techniques". They introduce the idea of components: objects that encapsulate both structural and geometrical data, as well as optimized algorithms for the manipulation and representation of the data. These components are then linked together as nodes in a directed graph that describes the overall structure of the plant.

Lintermann and Deussen describe three types of component: components for generating geometry, components for the iteration and arrangement of other components and components that implement constraints and geometrical transformations. These components are briefly introduced below.

Geometry-generating components include:
Simple
Produces simple geometry such as cubes and spheres
Revo
Produces a surface of revolution
Horn
Used to generate stem-type geometry
Leaf
Produces leaf geometry

Iterating components include:
Tree
Distributes subsequent components as branches about a stem
Hydra
Distributes subsequent components in a radial fashion, perpendicular to its parent component
Wreath
Similar to the hydra except that subsequent components are distributed parallel to its parent component
PhiBall
Distributes subsequent components on the section of a sphere

Geometrical transformation and constraint components include:
World
Allows environmental elements such as light and gravity to affect subsequent components
Pruning
Constrains subsequent geometry through a pruning algorithm
Free-form-deformation
Allows manipulation of the overall plant structure by means of a 2D grid of control points
HyperPatch
Similar to free-form-deformation, except that a 3D grid of control points is used.

6.3.1   Practicality

Lintermann and Deussen show that their component modeling system is indeed simple and intuitive by means of a test in which inexperienced users were required to model complex plant structures with only a brief introduction to the system. The results produced by the inexperienced users were impressive (see Figure 6.3), demonstrating the ease-of-modeling provided by the system. Their system also includes a powerful user interface, through which plant components can be quickly and easily modeled and linked to form the overall plant structure.

Figure 6.3: Models produced by inexperienced users using a component modeling system

6.3.2   Realism

The degree of realism of this technique lies heavily in the hands of the user. Through manipulation of the component parameters and component hierarchy, both realistic and non-realistic looking plant models can be achieved. The models demonstrated by Lintermann and Deussen, while not necessarily being botanically accurate, show a strong resemblance to the actual plants they were intended to mimic.

6.3.3   Real-time Rendering

Lintermann and Deussen have given a fair amount of thought to the speed of rendering of the models produced. Many of the models generated by their system required upward of one second per frame, rendering on an SGI Indigo 2 Extreme (134 MHz).

Lintermann and Deussen suggest two techniques to improve poor rendering times:
- **Complexity reduction**
  Reducing the number of triangles displayed thus increasing the rendering speed
- **Hiding of components**
  A hidden component produces no geometry but generates subsequent components in the usual manner

The first is the most important, and widely used, technique for achieving rendering speed-up. The increased rendering speed obtained by a reduction in the number of polygons of a model often outweighs the loss of detail, and hence realism. Small plant models with few components are able to make use of complex geometry while maintaining suitable rendering speeds, while large plant models often need to sacrifice geometric complexity.

6.4   An Object-Oriented Approach To Modeling

G. Tankard developed a system using an object-oriented approach [45]. His system takes a more abstract approach to component modeling, providing graphical interface, instead requiring the modeler to preset the parameters of each component and set the associations between the components. The practicality lost through the exclusion of the user interface is offset somewhat by the improved modeling precision attained.

Because of the object-oriented framework for modeling, rather than a fully-fledged modeling application, the system is designed to be highly extendable. Easy creation and integration of new components into the modeling framework are done. Some of the components that are essential to the system are as follows.

- Stem component (Horn)
- Leaf component (Leaf)
- Tree component (Tree)
- Radial component (Hydra)

Design of the Component Structure
Components are intended to encapsulate data (attributes) as well as algorithms for the manipulation and representation of that data. It is for this reason that a purely object-oriented approach should be chosen. The encapsulation of data and corresponding methods is an inherently object-oriented idea. In addition to these, all components have the following common properties.

A set of component connectors
All objects have the ability to have other components connected to them as children:

- Each component is responsible for its own visualization
- Each component contains algorithms for the visual representation of its data
- Each component, after visualizing itself, instructs its children to do the same

Attributes
Two techniques are discussed here for the specification of plant attributes.

- **Single-valued attributes**
  Plant parameters often require only a single value in their definition. Examples of this, in the context of component modeling, are the amount of geometric detail required and the number of branches a tree should generate.

- **Spline-valued attributes**
  Inspired by the research of Prusinkiewicz et al, into the "use of positional information in the modeling of plants", this method of plant-attribute specification involves the definition of a spline curve that allows a given attribute to vary over distance.

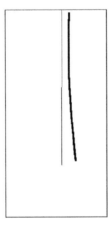

Figure 6.4: A spline definition of the stem radius

Examples of attributes that are suited to spline definition are the shape and radius of a stem. Figures 6.4 show the spline definition of stem radius. It is the distance between the spline and the line at that particular point. Figure 6.5 shows the resulting geometry.

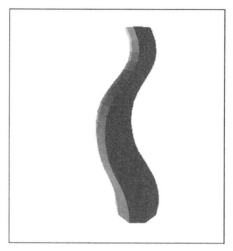

Figure 6.5: The resultant geometry

The components introduced in this section are those that generate geometry. The purpose of each component as well as various ways in which the component geometry can be created are now discussed.

6.4.1 Stem Component

This component generates cylindrical geometry for the representation of stems, branches, twigs and any other plant element with similar geometric properties. There are a number of ways in which stem geometry can be produced. Two of these techniques are introduced below.

Cylinders and cones
A cylinder is a "solid or hollow body with straight sides and a uniform cross-section". Most stems and branches are, to an approximation, cylindrical in nature making cylinders a natural choice in the representation of stem-type geometry. Not only are cylinders trivial to model, their low complexity also makes them fast to render.

These advantages are offset by the fact that cylinders, by nature, are without curvature, and while this can be a fair approximation, few stems or branches on plants are actually straight. Another problem when using cylinders for stem geometry occurs at the point where two branches meet. There is no smooth transition from one branch to another, causing the plant to have an unnaturally angular appearance and also causing "holes" to appear in the model as demonstrated in Figure 6.6.

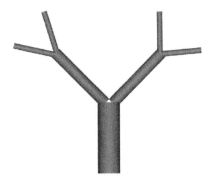

Figure 6.6: Problems observed when modeling with cylinders. Holes, such as the one above the trunk in this image, can appear in the mesh detracting from the realism of the model.

A possible solution to this problem is to use cones. The only difference between a cone and a cylinder lies in the fact that cones have different start- and end-radii, thus causing them to appear less uniformly straight. Another possible solution is to apply texture mapping to the cylinder or cone as this often gives the appearance of smoother geometry.

Generalized cylinders

A vast improvement on regular cylinders and cones, generalized cylinders overcome many of the problems introduced above. A generalized cylinder can follow any path and can have a varying radius along its length. They allow for the creation of smoothly curving stems and branches and, if modeled correctly, allow smooth transitions between branches and prevent holes from appearing at branching nodes.

The only major disadvantage of generalized cylinders is their complexity. Due to their curvature, generalized cylinders require many times more polygons than regular cylinders. This is demonstrated in Figure 6.7. They also require significantly more computation to generate, due to the number of mathematical operations needed to calculate vertex positions.

Figure 6.7: A generalized cylinder

The benefits of generalized cylinders greatly outweigh their shortcomings. The added realism produced by non-rigid stems and branches is well worth the sacrifice of having higher complexity.

The function of the Stem component is to generate geometry for plant elements such as stems, branches and twigs, the geometry that is cylindrical in nature. Some important attributes of the Stem component are listed in Table 6.2 below.

Attr. Name	Description
stemShape	A spline defining the shape of the generalised cylinder used to represent the stem geometry.
stemRadius	A spline defining the radius of the stem. Only the x-coordinates of the spline are used in calculation of the radius.
detail	The number of stemShape spline iterations used – a high detail value yields a very smooth curve.
meshDetail	The number of vertices used per spline iteration – the number of sides in the resulting cylindrical shape.

Table 6.2: Attributes of the Stem component

The Stem component produces as its geometry a form of generalized cylinder that is a dynamic form of cylinder that follows a spline-defined path and has a radius that varies along its length. The generalized cylinder is constructed in the following manner.

A number of equally spaced points, together with their orientations, are obtained from the stemShape spline. For every pair of points obtained:
  A circle of vertices is generated about each point.
  • The circle lies in a plane perpendicular to the tangent vector of a point.
  • The number of vertices is specified by the *meshDetail* attribute.

For each pair of points, quadrilateral polygons are generated linking the corresponding vertices

This process is demonstrated in Figure 6.8 below.

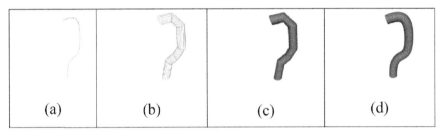

| (a) | (b) | (c) | (d) |

Figure 6.8: Construction of a generalized cylinder

At each point on the stemShape spline, a ring of vertices are calculated in the x-z plane using the following equations:

$$x = r \cos\left(i * \frac{2\pi}{meshDetail}\right)$$

$$z = r \sin\left(i * \frac{2\pi}{meshDetail}\right)$$

where r is the radius of the generalized cylinder at that point on the stemShape spline, and i is an integer ranging from 0 to (*meshDetail*-1). The inclusion of the factor 2π is due to the fact that the angles are measured in radians.

Once these vertex-positions have been calculated, each vertex is rotated about the z-axis by the angle of the tangent to the stemShape spline at that point. Finally, the vertices are repositioned so as to be centred about the stemShape spline point.

6.4.2   Leaf Component

The leaf component is responsible for the generation of leaf-type geometry. This includes plant elements such as petals and even grass.

These elements are two-dimensional by nature – a fact that can be taken advantage of for the purposes of practical modeling. Two methods of modeling leaves are discussed here.

Polygon definition

Defining leaf geometry by polygon definition involves the explicit definition of each polygon point, or vertex, of the leaf outline. The advantage of this technique is the ability to model the leaf shape exactly, allowing for the creation of any types of leaf.

The primary disadvantage of this technique is its inability to model smooth leaves accurately. Since each vertex in the leaf shape must be explicitly specified, a smooth curve would require the definition of many vertices – a highly impractical approach.

Spline definition

Using a spline to define the outline of the leaf alleviates some of the problems introduced in the polygon definition method above. Smooth curves are easily modeled with splines using only a few strategically placed control points.

The major disadvantage of this technique is the inability of a single spline to model jagged edges. Leaves seldom have smooth edges all around. Most at least have a pointed edge.

A partial solution to this problem is to model only one half of the leaf. Due to the symmetry of leaves in general, the other half of the leaf can be assumed to be a mirror image of the first. This allows for the modeling of a pointed edge as shown in Figure 6.8, but still does not allow for jagged leaf edges.

Either method of leaf-geometry generation is acceptable. The arguments for each of these two methods is based on potentially enhanced realism (polygon definition) versus ease of modeling (spline definition).

Figure 6.8: Through spline definition of one half of a leaf, the other half can be mirrored, allowing for the creation of pointed leaves.

The function of the class LeafComponent is to generate leaf-type geometry. The primary attributes of LeafComponent are listed in Table 6.3 below.

Attr. Name	Description
leafShape	A spline that defines the outline of the leaf
detail	The number of leafShape iterations used

Table 6.3: Attributes of the Leaf component

As mentioned in Table 6.3, leafShape defines the outline of the leaf. Due to the symmetrical nature of leaves, only one half of the outline of the leaf can be defined. The other half can be calculated by mirroring the spline coordinates about the line x=0. The process of leaf geometry generation can be summarized as follows:

- A number (detail) of equally spaced points are extracted from the leafShape spline
- For each point, now treated as a vertex, a corresponding vertex is created with the same y-coordinate but a negated x-coordinate
- These vertices are used to describe a polygon of the colour specified by r, g and b.

6.4.3   Tree Component

The Tree component can be categorized as distribution component that generates no geometry directly. It has a Stem component attribute that may generate geometry, but for the purposes of this discussion this will not be considered a direct creation of geometry. The purpose of a distribution component is to generate multiple instances of subsequent components and arrange them in a specified manner about a parent component.

The Tree component provides for the distribution of subsequent components as branches about a central axis, defined by a Stem component. The Stem component is not only used to calculate the positions and orientations of the Tree branches, but can also be used to generate the geometry of the Tree stem. The distribution properties of the Tree component are introduced here.

Positioning of branches
There are a variety of ways in which branches can be positioned along the stem. One method is to define the distance between each branch. This can be a constant value, in which case branches are equally spaced along the stem or a spline.

A simple method of positioning branches is to specify the number of branches required and spread the branches equally along the stem. This is identical to defining a constant-valued distance between branches, except that a specification of the number of branches required is more intuitive.

Orientation of branches
Once the positioning of the branches are determined, the next step is to determine the orientation of the branches. Two orientation properties are used in the Tree component and both will be discussed here and are demonstrated in Figures 6.9 and 6.10.

Branching angle
The branching angle is the angle that the branch makes with the stem. This can be accomplished through the use of a spline, but this proves clumsy without the use of a graphical user interface with interactive feedback. The branching angle can be the same for all branches. However, this can detract from the realism of the plant as all stems have branches that have the same branching angle. Hence, an attribute namely *branchAngleVariation* can be introduced here, which causes a specified amount of deviation from the values of their parents.

$$a_v = a + \left(v_a - 2rv_a\right)$$

where a_v is the varied branch angle, a is the unmodified branch angle, v_a is the *branchAngleVariation* and r is a random number between 0 and 1.

Inter-node angle (Phyllotaxis)
The inter-node angle, or phyllotaxis, of a plant specifies the angle about the stem axis between two successive branches. In a majority of plants, this angle is fairly constant. While a random angle will provide realism, it requires graphical user interface for the purpose.

Figure 6.9: Branching Angle   Figure 6.10: Inter-node Angle

Scaling of branches
Very seldom in a tree structure are the branches all the same size. The size of the branches usually varies over the length of the stem in a pattern that differs from one plant to another; some plants have large branches at the base of the stem and small branches at the top, while others are exactly the opposite and still others have small branches at the top and bottom with large branches in the middle. For the sake of realism, the user can define the scaling function. The branch scaling spline effectively defines the shape of the Tree component.

An attribute namely *branchScaleVariation* can be introduced here, which causes a specified amount of deviation from the values of their parents. The variation attributes are specified as the maximum fraction of the value of their parents by which they can deviate. The equations used in the calculation of the variations are as follows.

$$s_v = s - 2rv_s s$$

where s_v is the varied scale, s is the unmodified scale value, v_s is the *branchScaleVariation* and r is a random number between 0 and 1.

6.4.4   Radial Component

The function of this component is to distribute subsequent components in a radial fashion about a centre point. Some of its most common uses are in the distribution of petals to form a flower or the distribution of leaves about a stem. By assigning Radial components as branches of a Tree component, the effect of multiple branches per "branch-node" can be achieved.

The distribution properties in Radial component affect only the orientation of subsequent components. As all subsequent components are distributed about a single centre point, there is no need for any positional properties as in the Tree component. The three basic properties that affect the distribution of components in the Radial component are as follows.

Inter-node angle
This is similar to phyllotaxis that specifies the angle, about the axis of the parent component, between nodes.

Node elevation angle
This is similar to branching angle that specifies the angle that each node makes with the axis of the parent component.

Node twist angle
This specifies the angle of rotation of a node about its own axis.

These parameters are fairly self-explanatory, and as such will not be discussed in any more detail here.

6.5   Other Tree Modeling

Prusenkiewicz *et al* introduced the "use of positional information in the modeling of plants". Based on L-Systems, this technique aims at the integration of artistic elements such as posture, arrangement of components and plant silhouette into plant models.

Jirasek *et al* introduced the idea of rotational springs in the modeling of the effects of gravity on plants. Each branch is divided into a number of segments connected by rotational springs. Each segment has length, width and mass properties, and each rotational spring is assigned an individual spring constant. Hanging plants are modeled particularly effectively with this method. Some of the other tree modeling techniques are given in the following sections.

6.5.1   Tree Growth Visualisation

Linsen et. al. [83] used equations derived from measured street tree data for London Plane tree (Platanus acerifolia) such as tree height, diameter-at-breast-height, crown height, crown diameter, and leaf area to model photo realistic images of many trees of the selected species. Models describing the fractal branching structure of trees typically exploit the modularity of tree structures. The models are based on local production rules, which are applied iteratively and simultaneously to create a complex branching system. The authors tried to visualize the growth of a prototypical tree of certain species with a realistic look and conforming the real, measured data.

A branch or internode is defined by its length l, diameter d, start point s and direction Δ. A bifurcation or node is defined by the angles ϕ_i between the axes of the parent branch and the child branches and by the ratios in length l_i / l_0 and diameter d_i / d_0 between the parent branch and the child branches, i = 1,2. When one of the child branches bifurcates again, it will, in general, not lie in the same plane but in a plane of different orientation. The change in orientation is defined by the divergence or twisting angle θ_i, i = 1,2. In addition to the nodes and internodes, there are leaves and flowers. No production rules are applied to leaves and flowers, but they can grow in size.

A parametric L-system was used to describe the tree model. The chosen parametric L-system was defined as a context-free or context-sensitive grammar G = (V,T,S,P), where the set of variables V consists of branches B(l,d,s,Δ) and the trunk T(l,d,s,Δ), the set of terminals T consists of leaves L(s), the start symbol is the trunk T(l,d,s,Δ) and a set of production rules as given below.

Trunk length and diameter

The length l and the diameter d of the trunk are directly controlled by the global functions. The length $l = l(t)$ is defined as the difference between the measured height of the tree and the measured height of the crown i. e.,

$l(t) = f_H(t) - f_{CH}(t)$

where $f_H(t)$ & $f_{CH}(t)$ are the resulting functions for height & crown height respectively for the London Plane trees.

The diameter $d = d(t)$ is directly proportional to the measured diameter-at-breast-height DBH i. e.,

$d(t) = c_{DBH} - f_{DBH}(t)$

where $c_{DBH} \in [1, 1+\varepsilon)$ for a small $\varepsilon > 0$.

Branch length

When a branch grows, it exhibits a similar growth rate as the trunk or the tree as a whole. Thus, the length of a branch follows the growth rates of the respective functions. To assure that the tree model has the actual, measured tree height, the function $f_H(t)$ is used to control the elongation of internodes.

Intuitively, primary branches (i.e., branches that emanate from the main branch/trunk) start growing before secondary branches (i.e., branches that emanate from primary branches) exist and so on. Thus, primary and secondary branches do not grow at the same rate, while primary branches may already have reached a slow-growing phase; the secondary branches may still be in their initial fast-growing phase. Thus the time of creation t_0 of a branch is maintained to keep track of this process and to compute the growth with respect to this point in time.

Moreover, a secondary branch does not reach the length of a primary branch and a tertiary branch does not reach the length of a secondary branch, etc. Therefore, the growth function is multiplied with a scaling coefficient c_l. The scaling coefficient c_l of a branch is obtained from the scaling coefficient of its parent branch multiplied by the scaling factor s_l $\in (0,1)$, where the trunk has a scaling coefficient c_l of value one. The scaling factor s_l depends on the species. For the London Plane tree, we use random values $s_l \in (0.6, 1)$. The randomness is required to make the tree appear less symmetric and thus more realistic. In summary, the length $l = l(t)$ of a branch at time t is given by

$l(t) = l_{max} / H_{max} \cdot c_l - f_H(t - t_0)$

where l_{max} and H_{max} are the maximum length of the branch and the maximum measured height of the tree respectively. The maximum length

l_{max} of a branch is determined by the maximum length of the parent branch multiplied by the scaling factor s_l. The growth of the branch terminates when the maximum length is reached.

Branch diameter

When a branch bifurcates, the child branches have a smaller diameter than the parent branch. Leonardo da Vinci postulated that the square of the parent's diameter is the sum of the squares of the diameters of the children. In a dynamic setting, we use the measured function $f_{DBH}(t)$ multiplied by a scaling coefficient c_d to determine the growth of the diameter $d = d(t)$.

The scaling coefficient c_d is based on the scaling coefficient c_d' of the parent branch but also on the scaling coefficient c_l, which establishes a correlation between the scaling in length and diameter. The scaling coefficient is computed as $c_d = c_l - c_d'.(1-0.7. c_d')$. The trunk has a scaling coefficient c_d of value one. Different diameters for different branches are induced by the randomness in the scaling coefficient c_l. The growth in diameter is computed with respect to the time of creation t_0. The diameter $d = d(t)$ is defined by
$$d(t) = c_d . f_{DBH}(t - t_0).$$

Branch orientation

The orientation of a branch is determined by the orientation of the parent branch, the bifurcation angle ϕ_f and the divergence angle θ. The bifurcation angle f is based on the ratio of crown diameter $f_{CD}(t)$ and crown height $f_{CH}(t)$, which defines the shape of the crown. London Plane trees are vertically ellipsoidal, which means that their crown height is greater than crown diameter. The ratio of crown diameter $f_{CD}(t)$ and crown height $f_{CH}(t)$ is approximately constant over time. The bifurcation angle is taken as
$$\phi = \arctan (f_{CD} / f_{CH}) \mp \alpha,$$
where α is a small random angle to make the tree less symmetric and thus more realistic.

When choosing divergence angles θ, it is required that the urban street trees regularly visualized and pruned to obtain an "optimal" shape. A balanced tree, where primary branches called scaffolds are evenly spaced radially around the trunk, is considered optimal. Also, lower branches are removed to allow for clearance by trucks. Therefore, the authors considered θ as $130°$ to have evenly spaced branches spiraling up the trunk.

6.5.2 *Adaptive L-system*

Shing et. al. discussed a simulation system named SimEco using an adaptive L-system to interact among the plant organs and environment by particle system by integrating the Particle System into the L-system. The basic concepts of SimEco are to describe the hierarchical botanic structure of plants by L-System and to perform the physical effects, which act on the plant organs by particle system. In the SimEco system, each organ of the Plant is treated as an independent particle generator. Each particle generator will shoot particle(s) according to the types of the organs. By calculating its trajectory hierarchically, the 3D position, orientation and other physical properties of each organ can be determined. This approach removes most of the parameters used in the traditional parametric L-system. The complexity of L-System grammar is therefore reduced substantially.

SimEco system combines non-deterministic and non-propagating L-system and particle system. When strings generated by L-System are fed to the rendering engine, each alphabet of strings will be treated as a particle generator. Each particle generator in Rendering Engine inherits the end position and end velocity of its parent particle. Generally speaking, each particle generator ejects only one particle. This particle then flies according to effects of environment, such as gravity or sunshine. In fact, the trajectory of particle is the skeleton of organs. SimEco system will set up models of organs in terms of the skeletons of organs.

SimEco system provides a script language named SimEco Script for user to describe the botanical properties of plants and environment. First, SimEco system will interpret the plant grammar, which is written by tools and users in order to obtain the properties of environment, the stages and the growing grammars of each plant defined in the garden. The system repeatedly iterated the plant grammars in terms of the growing grammars and the strings of plants generated by L-System possess the hieratical relationship among organs. L-System delivers strings step by step. When obtaining strings form L-System, Rendering Engine will calculate the actual 3D position of organs in terms of the interaction among influence of environment and particles.

6.5.3 *D0L System with 3D Structure*

Strings generated by L-systems may be interpreted geometrically in three-dimensional way. The *turtle interpretation* of L-systems, introduced by Szilard and Quinton and extended by Prusinkiewicz is presented earlier in section 2.3 of chapter 2. After a string has been generated by an

L-system, it is scanned sequentially from left to right, and the consecutive symbols are interpreted as commands that maneuver a LOGO-style turtle in three dimensions. The turtle is represented by its state, which consists of turtle position and orientation in the Cartesian coordinate system, as well as various attribute values, such as current color and line width. A vector P defines the turtle position and three vectors H, L and U, indicating the turtle's heading and the directions to the left and up as shown in Figure 6.11, define the orientation. These vectors have unit length, are perpendicular to each other, and satisfy the equation H x L = V.

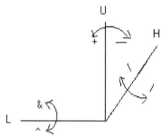

Figure 6.11: Turtle in Three Dimensions

The symbols that control turtle orientation in space as shown in the above figure are given as below.

+: Turn left by angle θ around the V axis.

-: Turn right by angle θ around the V axis.

&: Pitch down by angle θ around the L axis.

^: Pitch up by angle θ around the L axis.

/: Roll left by angle θ around the H axis.

\: Roll right by angle θ around the H axis.

|: Turn 180° around V axis. This is equivalent to +180° or −180°.

The symbols for modeling structures with branches are as follows.

[: Push the current state of the turtle (position, orientation and drawing attributes) onto a pushdown stack.

]: Pop a state from the stack and make it the current state of the turtle. No line is drawn, although in general the position and orientation of the turtle are changed.

These symbols are used to construct three-dimensional trees. For this purpose the following set of rules and axioms under D0L are written.

3DTree{

Dirs = 3

Axiom = F

F=F[-F[-F]F]/F[-F]F

};

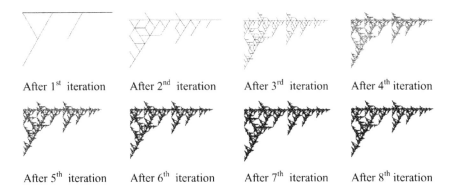

| After 1st iteration | After 2nd iteration | After 3rd iteration | After 4th iteration |

Figure 6.12: Shape of 3D Fractal trees after various iteration

The images after various iterations are shown in the figure above.

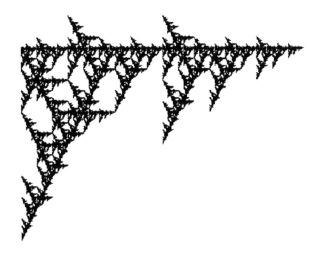

Figure 6.13: 3D Fractal trees

A stable fractal image is generated after the 7th iteration. The 3D views of the trees may be seen in the above figure 6.13.

6.5.4   *Environment Interactive*

The environment friendly functionalities can generate a more realistic tree. By interacting with its environment, a plant may be pruned with respect to a surface. This enables the designer to determine the final shape of the plant more easily and to generate artistic and realistic plants. An interpreter can be generated with the capability of handling biological development phase of a tree using parametric, conditional, context sensitive and environmentally sensitive productions. These properties of the interpreter models plants realistically. The following three productions are depicted below.

- Parametric and conditional productions are required for allowing the nodes to elongate, and for making the plant keep track of time.
- Context sensitive productions are required for simulating the hormone system of the plant.
- Environmentally sensitive productions enable the plant to interact with its environment, also they enable us to create synthetically structured plants.

When a plant is pruned, it responds by producing new branches. During the normal development of a tree, buds are created, but many buds do not produce new branches, and remain dormant. These buds may be subsequently activated by the removal of leading buds from the branch system. This event is called traumatic reiteration. It results in an environmentally adjusted structure. The following L-system represents the extreme case of this process, where buds are activated only as a result of pruning.

$w : FAP(x,y,z)$
$p1: A > P(x,y,z) : !prune \rightarrow BF/(180)A$
$p2: A > P(x,y,z) : prune \rightarrow T$
$p3: F > T \rightarrow S$
$p4: F > S \rightarrow SF$
$p5: S \rightarrow E$
$p6: B > S \rightarrow [+FAP(x,y,z)]$

A user-defined function prune is defined below that is either true or false depending on the current position coordinates, x, y and z. The axiom w initializes the plant with an internode F, an apex A and a query module $P(x,y,z)$. Initial development is performed by the production p1. At each application of this rule, a dormant bud B is created, and the plant grows more. The module $/(180)$ rotates the turtle around its own axis. When the plant is to be pruned, T replaces P by the production p2. Then p3 converts T into bud-activating signal S. Then S goes basipetally (downwards) via the productions p4 and p5. Production p6 induces a dormant bud and makes it initiate a lateral branch. The process of creating the activation signal and activating a dormant bud process repeats itself until all dormant buds are activated.

A prune function can defined as:
$f(x) = $ true if $-0.5 < x < 0.5$, $-1 < y < 1$, and $-0.5 < z < 0.5$
$f(x) = $ false otherwise

Parametric productions are the ones in which the left-hand side of the production has a parameter and the newly created tokens that are on the right-hand side have parameters depending on the parameter of the left-hand side. These productions are in the following form:

$R(t) \rightarrow L_1(f_1(t)) L_2(f_2(t)) \ldots L_n(f_n(t))$

In these productions, functions applied on **t** do not contain very complicated operations. Increment decrement or multiply operations are enough for the program's purposes. The production token objects have type and operand parameters, which enables us in writing parametric productions. The following functions can be used when calculating $f(t)$.

- Constant function: $f(t) = 0$
- Decrement function: $f(t) = t - x$
- Increment function: $f(t) = t + x$
- Multiply function: $f(t) = t \times x$
- Initialize function: $f(t) = x$
- Equal function: $f(t) = t$

As there is an ambiguity in the parameters of the tokens of the parametric L-systems, where the parameter of a node can be used for two purposes, as follows.

- The parameter can be used for representing the length of the node, where at each iteration, the value of the parameter can be incremented, which makes the older nodes become more developed from the younger ones.
- The parameter can also be used for timekeeping, where a segment or a signal is produced after waiting a certain amount of time. In this case, parameter has no effect on the length of the node.

For differentiating between these cases, an additional parameter called Special Parameter can be added. If the special parameter is a positive number, then the parameter of the node represents its length and the initial length is equal to special parameter. If the special parameter is zero, then the parameter of the node does not represent the node's length; it may be used for timekeeping.

Besides applying the above parametric productions, sometimes conditional productions are applied in order have certain conditions. For example, the following productions may be used, if the growth of internodes is to be stopped, when their length reaches 1.2 units.

p1 : Internode(i) : $i \neq 1.2 \rightarrow$ Internode (i+0.2)
p2 : Internode(i) : $i = 1.2 \rightarrow$ Internode (1.2)

When 'i' is equal to 1.2, p2 is applied; otherwise p1 is applied.

In order to make the production rules environmentally sensitive that is to allow the plant to interact with its environment, pruning is implemented. In order to implement pruning, each node of the plant should know its position coordinates. Then the following algorithm can be implemented.

repeat

Step 1: Calculate coordinate values for each node.

Step 2: Iterate the plant according to the L-system rules considering the coordinate values.

until done

While applying the rules, the program considers the coordinate values of a node before applying a production on it. Not all of the productions have to depend on the coordinate value; some may be applicable whatever the position is.

For the productions that depend on the coordinate values, the following convention may be implemented: If a node is to be pruned to a surface, then provide two productions, one for the case "if the node is inside the surface", and other for the case "if the node is outside the surface". Then the program performs the operation

Applicable ← (Is Node Inside) XOR (NOT Apply rule to prune)

in order to determine if a production is applicable or not. Note that, if two productions with the same token and the same inside function are given, with one having "Apply to prune" as true, other having it as false, then we are guaranteed to find an applicable production for every point in space.

In order to make a node disappear after being pruned, we have to add a rule like $X \rightarrow \varepsilon$, where ε denotes the empty string. However Context Sensitive Grammars in the Chomsky Hierarchy do not allow such productions to be present since length of left hand-side is greater than the length of the right hand-side. As a solution, the productions requiring empty string on the left are written as: $X \rightarrow$ dummy, where "dummy" is a node with zero length and with no productions defined which transform it to another node.

Production format

A production consists of three steps, writing left-hand side, writing right-hand side and placing them into a data structure. The left-hand side of the production will have the following.

- Left context (Predecessor)
- Right context (Successor)
- Production type (parametric, conditional, context sensitive or environmentally sensitive)

On the right-hand side, tokens that replace the token on the left-hand side are written. Integer identifiers are generally given to the tokens, while writing the right-hand side. Additionally there are special tokens ROTX,

ROTY, ROTZ, LSQBCKT, RSQBCKT, LEAF and FLOWER that make the graphical part act accordingly.

For finding an applicable production the following algorithm is performed.

Step 1: Get array of productions

Step 2: for each element in the array do

begin

Step 3: if production is context sensitive, control successor and predecessor

Step 4: if production is environmentally sensitive, check coordinates and equation

Step 5: if production is conditional, check if condition is satisfied

Step 6: break the loop in the first production that passes all checks

end.

For drawing the plants, the interpreter should describe the L-strings that it had generated to the graphics part, so that the plant can be visualized. For this purpose, after creating the final L-string, the interpreter makes a pass over the L-string and simulates the turtle graphics, while creating the necessary visual components. An algorithm can be performed in this step as follows.

    for each token in the L-string do
    begin
        if current token is [
            Push transformation matrix to the stack
        else if current token is]
            Pop transformation matrix from the stack
        else if current token is ROTX
            Tell graphics part to rotate around the x-axis by current token's
            parameter degrees
        else if current token is ROTY
            Tell graphics part to rotate around the y-axis by current token's
            parameter degrees
        else if current token is ROTZ
            Tell graphics part to rotate around the z-axis by current token's
            parameter degrees
        else if current token is LEAF
        begin
            set leaf's parameters
            Request the leaf to be displayed
        end
        else if current node is a visible internode
        begin

        Set its parameters
        Request it to be displayed
        Move forward by the length of the displayed node
    end
end

Leaf

An L-system can be written for representing the leaf. A realistic L-system for this purpose is given below:

$w : A(0)$

$p2 : A(t) \rightarrow G(LA) \; [\; -B \; (t) \;] \; [\; A(t+1) \;] \; [\; + B(t) \;]$

$p2 : B(t) \rightarrow H(LB) \; B(t-PD)$

$p3 : B(0) \rightarrow H(1)$

$p4 : G(t) \rightarrow G(t \times RA)$

$p5 : H(t) \rightarrow H(t \times RB)$

The parameters defined above are as follows.

- LA: initial length of main segment.
- RA: growth rate of main segment.
- LB: initial length of lateral segment.
- RB: growth rate of lateral segment.
- PD: growth potential decrement

It is possible to obtain different shapes of leaves by modifying these parameters using the parametric L-system. Many more different L-system codes may be created for different types of leaves [100].

Here according to production $p1$, in each derivation step, apex $A(t)$ extends the main leaf axis by segment $G(LA, RA)$ and creates a pair of lateral apices $B(t)$. New lateral segments are added by production $p2$. Parameter t, assigned to apices B by production $p1$, plays the role of "growth potential" of the branches. It is decremented in each derivation step by a constant PD, and stops production of new lateral segments upon reaching 0. Segment elongation is captured by productions $p4$ and $p5$. The number of lateral segments is determined by the initial value of growth potential t and constant PD. Since the initial value of t assigned to apices B increases towards the leaf apex, the consecutive branches contain more and more segments. On the other hand, branches in the apical area exist for too short a time to reach their limit length. Thus while traversing the leaf from the base towards the apex, the actual number of segments in a branch first increases, then decreases. As a result of these opposite tendencies, the leaf reaches its maximum width near central part of the blade.

The major disadvantage of this method is its storage cost and high structural complexity. Each leaf is represented as a tree, which causes the number of nodes increase a considerable amount. Additionally, at each iteration, program has to apply production rules for each one of the leaves, which is a time consuming job.

6.6   Summary

L-systems can be used to generate realistic plants in three dimensions by applying a turtle interpretation to the symbols of the grammar. Many surveys have been done for computer representation of trees for realistic and efficient rendering. However, this chapter includes techniques and procedure involving L-systems and any processes added to it for designing and developing realistic 3D trees.

Chapter 7

Fractal Dimension

In this chapter the calculation of fractal dimesion for various fractals along with those generated through L-system are discussed. Mandelbrot introduced the term Fractal having fractional dimension unlike dimensions in Euclidean geometry [21-25]. Pentland [101,102] provided the first theory in this respect by stating that fractal dimension correlates quite well with human perception of smoothness versus roughness of surfaces, with fractal dimension of 2 corresponding to smooth surfaces and fractal dimension of 3 corresponding to a maximum rough "salt-and-pepper" surface. He assumed the surface to be modeled by fractional Brownian function. However, it was followed by many other theories applicable to a wider class of fractals. Gangepain et. al. [67] described the popular reticular cell counting method and Keller et. al. [69,70,120,121] gave even more interesting theories. Sarkar et. al. [94] gave another method to find out the fractal dimension known as Differential Box Counting (DBC) method. Many more studies are made by various researchers in this direction [71,91,97,99,119,133,136]. Many studies are made to find the upper and lower bound of the box size and in this regard Keller et. al. [120] have given a theoretical justification for a restriction on the smallest box size inspired by the work of Pickover and Khorasani [28]. Lin et. al. [65] pointed out that under counting will occur in small scales and overcounting will occur in large scales in case there are uniform intensity areas. Mishra et. al. [4] found out a lower bound of the box size, which gave more accurate results.

7.1 Self Similar Fractals

The Hausdorff-Besicovitch dimension [122] and the related self-similarity dimension are used to define a fractal dimension. Fractal dimension of many artificial and natural objects cannot be determined analytically [26,27]. These can be calculated using compass and box dimension. We can even use a box counting to measure fractal dimensions of strictly self-similar fractals. A comparison between theoretical dimensions of such fractals with grid dimensions done by Cretzburg et. al. [113] shows a close match between them.

7.1.1 Fractals and Self-similarity

The Hausdorff metric **h** measures the distance between sets $A, B \subset X$ as

$h(A, B) = \sup \{d(x,B), d(y, A) \mid x \in A, y \in B\}$

The point $x \in A$ farthest from any point of B is found. Then the point $y \in B$ farthest from any point of A is found. The larger of these two maximal distances determines the Hausdorff distance between A and B. The d-dimensional Hausdorff measure $\mathbf{H}^d(A)$ of a set A is

$$H^d(A) = \lim_{\delta \to 0} \inf \sum_{i=1}^{\infty} diam \ (U_i)^d$$

where $\{U_i\}$ is a countable delta-cover of A.

The Hausdorff dimension **dim A** of set A is

$\dim A = \sup\{d : \mathbf{H}^d(A) = 0\}$

Also

$\dim A = \inf\{d : \mathbf{H}^d(A) = 0\}$

This definition was given by Hausdorff in 1919 and later generalised by Besicovitch in 1928.

An IFS [23] is a finite collection $w = w_i$ of contractions $R_n \to R_n$. For any IFS [23] $w = w_i$ there exists a unique non-empty compact set $A \in R_n$ such that $A = w (A)$. If A is the attractor of an IFS satisfying the open set condition and with Lipschitz constants S_i, i=1 ... N, then its self similarity dimension $\dim_s A$ is given by the solution **d** of

$1 = \Sigma s_i^d$

If all similarities share the same Lipschitz constant S then the self-similarity dimension **dim$_s$** is simply

$\dim_s = \log N / \log (1/s)$

Fractal can be defined as a set where the Hausdorff Besicovitch dimension exceeds its topological dimension [22]. Also a fractal is a shape made up of parts similar to the whole object [22], which shows the property of self-similarity. However, a self-similar object may not be a fractal. For example, a line is made up of two lines each shrunk by a reduction factor of 1/2. Hence it is self-similar. So the dimension $\dim_s = \log(2)/\log[1/(12)] = \log(2)/\log(2) = 1$. It is the Euclidean dimension. Similarly, a square is made up of four squares; each shrunk by a reduction factor of 1/2. Hence $\dim_s = \log(4)/\log[1/(1/2)] = \log(4)/\log(2) = 2$. If the similarity dimension is not Euclidean, then it is a fractional one and hence is called as fractal dimension. If we reduce any image by ½ as shown in figure 7.1 and place it in 1, 2 & 3 positions as shown in the blueprint in figure 7.2, a fractal known as Sierpinski (Figure 7.3) is generated. The Sierpinski's gasket is made up of 3 identical pieces, each shrunken from the original by a factor of 1/2. Hence the dimension $\dim_s = \log(3)/\log[1/(1/2)] = \log(3)/\log(2) = 1.584962$.

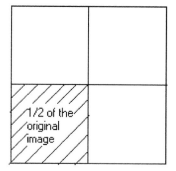

Figure 7.1: Reduction of image by 1/2

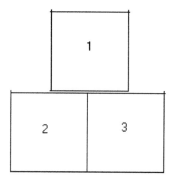

Figure 7.2: Blueprint

Figure 7.3: Sierpinski Gasket

7.1.2 Length of the coast of Iceland

What is the length of the coast of Iceland ? This can be calculated using both Compass and Box-counting methods. However, it cannot be a well defined length. In that case, a log-log plot between various measures r and length L has to be made. The Compass dimension **D** can be found out using the following formula.

$$L(r) = const. \cdot r^{1-D}$$

r (k.m.)	N	L = N.r	ln r	ln L
200	6	1200	5.30	7.09
100	14	1400	4.61	7.24
80	20	1600	4.38	7.38
40	46	1840	3.69	7.52
20	112	2240	3.00	7.71

Table 7.1: Length of the coast Iceland.

Applying above equation and ploting the log-log D is found out to be 1.27. Thus for a man putting fence at a distance of 0.01 k.m., L(0.01) can be found out to be $e^{8.5}.0.01^{-0.27} \cong 17000$ k.m. But for a man putting lighthouses at a distance of 50 k.m., L(50) can be found out to be $e^{8.5}.50^{-0.27} \cong 1700$ k.m. The coast length differs by a whopping factor of 10.

If we want to find out the length using box counting method, we can follow the formula given below.

$$N(r) = const. \cdot r^{1-D}$$

We can make grids of various sizes on the coastline and using the log-log of N vs. r, found out D to be 1.26 which closely matches with the compass dimension of 1.27.

7.2 Extending To Fractal Surfaces

The introduction of fractal dimension estimation by Mandelbrot was extended to calculate dimension of fractal surfaces. Pentland assumed the surface to be modeled by fractional Brownian function and calculated the dimension of the surfaces.

7.2.1 Mandelbrot's formula

'Fractal' is the terms invented by Mandelbrot to describe the shape and appearance of objects, which have the properties of self-similarity and scale invariance. He brought the term fractal from the Latin word fractus, which means irregular segments. There have been a wide variety of measures and dimensions assigned to the fractal sets. These dimensions may be based on the definition of a probability measure over a space of functions or the statistical measure over a range of scales. Mandelbrot first described an approach to calculate FD while estimating the length of a coastline.

If a line of length N mm is measured by dividers set at 1 mm, the number of steps is obviously just N. But if the points are set to distance k mm, the required number of steps **N'** will be

$N' = N / K = N \times K^{-1}$

Suppose, there is flat surface which is covered by M small squares of side 1 mm and the small squares are changed to be having sides of k mm then the new number of squares **M'** which covers the entire area will be

$M' = M / K^2 = M \times K^{-2}$

A similar argument can be extended to volumes and then the generalized equation for E dimensional Euclidean space will be

$U' = U \times K^{-E}$

where **U'** and **U** are the numbers of units of measurement for units size k and 1 respectively in Euclidean dimension E.

Suppose we have access to a length measure, **L** (size 1 units) from a straight line; the number of units size k that is required is similarly L/K. We can therefore write a highly redundant expression as

$L = L \times K^{1-E}$

where E=1, implies that the length measure is independent of the ruler length.

Now, for an irregular curve like a coastline the measured length does depend on the length of the ruler and hence the above equation can be rewritten as

$L_k = L_1 \times k^{1-D}$

where L_1 is the length obtained for site units,

L_k is the length for site k units

and D is a non integral value called as the fractal dimension of the curve.

As A_1 and D are constants for a specific coastline, D can be calculated by having a least square linear fit of the log-log plot of A_k and k. If m is the slope of the fitted line, then FD of the coastline will be 1-m. Thus m should be always negative.

This same idea can be extended to the area of rough surfaces such as landscapes. Hence rewriting the above equation for surfaces, we can get

$A_k = A_1 \times K^{2-D}$

7.2.2 Casey's Classification

Casey et.al. [122] employed the Hutchinson dimension and gave emphasise on the estimation of contraction constant on self-similar sets only. Hutchinson dimension $D_{HB}(X)$ can be given by the solution to the equation $\sum \alpha_i^D = 1$, where $\{\alpha_i\}$ is the set of contraction constants and X be a self-similar set defined by contraction mapping m_i, $i = 1,2,3\ldots\ldots.n$. Casey et.al. explained that for $0 < D_{HB} < 1$, we will approximate a type of

Cantor set. If $D_{HB} = 1$, then we will produce a curve, For $1 < D_{HB} < 2$, we could approximate a curve on set of curve with fractal properties, similar to Koch curve or Sierpinski gasket. If $D_{HB} = 2$, it will be space filling. If $D > 2$, the speed generates filled up regions, "overlapping" many points $D/2$ times. This in fact results in the fractal surfaces.

7.2.3 Pentland's Method

Pentland [101,102] suggested a method of estimating FD by using displacement vector image intensity surface. Assume that the intensity $\mathbf{I}$ of a square image of N by N picture elements is given by

$\mathbf{I} = \mathbf{I}(x, y), 0 \leq x, y \leq N - 1$

Define a displacement vector ω as $\omega = (\Delta x, \Delta y)$, Δx, Δy are integers. Again define the difference of the image intensity at point (x, y) for a specific displacement vector ω as ΔI_w given by

$\Delta I_w(x, y) = \Delta I_w = \mathbf{I}(x, y) - \mathbf{I}(x + \Delta x, y + \Delta y)$

The above equation gives the difference of the image intensity of a picture along a specific displacement vector ω, whose beginning and end points are (x, y) and $(x + \Delta x, y + \Delta y)$ respectively. Pentland used the mean of the modulus of the difference image as defined in the above equation to estimate the statistical behavior of a texture at any specific scale as a function of displacement vector length in a constant direction. As fractal objects are self-similar and scale invariant, Pentland argued that a relation analogous to the one relating to fractional Brownian Motion (FBM) might be valid for image intensity differences. If $X_r(t)$ is the r-sampled random position of a particle at the time t, which is determined at the origin by $X_r(0) = 0$ with a probability 1, with a probability function for $X_r(t + \Delta t) - X_r(t)$, normally distributed with variance Δt, the probability of having $X_r(t + \Delta t) - X_r(t) \leq x$ for $t > 0$ and $h > 0$ is

$$P(X_r(t + \Delta t) - X_r(t) \leq x) = (2\pi\Delta t)^{-1/2} \int_{-\infty}^{x} \exp\left(-u^2/h^{2a}\right) du$$

The FBM model of index H follows a probability distribution similar to the above equation but the variance is dependent on the index H such as

$$P(X_r(t + \Delta t) - X_r(t) \leq x) = (2\pi)^{-1/2} \Delta t^{-H} \int_{-\infty}^{x} \exp\left(-u^2/h^{2a}\right) du$$

It is implicit in the above equation that the increments are stationary. However the distribution of the functions of the FBM cannot have independent increments. In fact, the expectation for these increments estimated with that probability function gives:

$$E((X_f(t + \Delta t) - X_f(t))^2 = \Delta t^{2H}$$
i.e $E(\Delta y^2) \propto \Delta t^{2H}$
i.e $E(|\Delta y|) = k \Delta t^H$

Pentland's argument was that under certain assumptions of illumination and surface reflectance's an image may be fractal and should have the following analogous spatial property

$$E(|\Delta I_\omega|) \times |\omega|^{n-1} = E(|\Delta I_{ref}|) = constant$$

where $E(|\Delta I_{ref}|)$ is a reference value, chosen for a displacement vector ω of unit magnitude. Pentland constructed the difference images of natural textures and plotted $E(|\Delta I_\omega|)$ as a function of scale, which he took as the absolute value of the displacement vector ω. According to him if the standard deviation of the image intensity differences is plotted against absolute element distance on a log-log scale, an estimate of the fractal dimension can be calculated from the slope of the resultant curve, which should be a straight line in case of a genuine fractal. Here the fractal dimension D of $X_t(t)$ is related to H as

$$D = 2 - H$$

Further, $X_t(t)$ has a random-phase fourier spectrum with power P(t) such that

$$P(f) \propto f^\beta$$

H is related to β by

$$\beta = 2H + 1$$

From the least square fit of the log-log of P(f) and f, one can estimate FD of an image intensity surface. Pentland also used this alternative method to calculate FD. But in either case one has to first model the image intensity surface to be a fractal Brownian function. However Pentland confirmed that the image of fractal surface is also fractal [2].

7.3. Various Methods Of Fractal Dimension Calculation

From the properties of self-similarity fractal dimension **D** of a set A in defined as

$$D = \log(N) / \log(1 / r)$$

Where N is the total number of distinct copies similar to A and A is scaled down by a ratio of $1 / r$.

7.3.1 Reticular cell counting method

In order to extend the above equation to the world of images, Gangepain and Roques-Carmes [67] introduced the cell counting method. If M × M is the size of an image and we imagine grids of size L × L, then we can cover up the entire image by boxes of sides L × L × L' in the vertical direction. Here L' = ⌊L × G / M⌋ can be a multiple of the gray level units where G represents the total number of gray levels, **N** is calculated by counting the total number of boxes that contain at least one gray level intensity surface. D is calculated by considering various L values, which in turn means various 1 / r values as 1 / r = M / L, M being constant. From the above equation, we have

$N \propto L^{-D}$

i.e. $N(L) \propto L^{-D}$

For each L, value of N is calculated and a log-log plot of N versus L is made. The slope of the least square linear fit line will be -D.

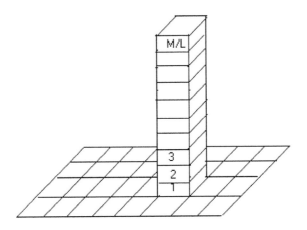

Figure 7.4: The boxes on a particular grid in the vertical direction

7.3.2 Keller's Approach

Voss showed that the graph of a FBM also satisfies this power law [115-116]. So Keller et. al. proposed a modification of the equation given by Roques et. al.. Let P(m, L) denote the probability that there are m intensity points within a box of size L centered about an arbitrary point of image intensity surface. Then

$$\sum_{m=1}^{N} P(m, L) = 1 \; \forall L$$

where N is the number of possible points in the box of side length L. If the image size is M×M, then the total points on the image will be M^2. Thus, the expected number of boxes with side length L needed to cover the whole image is given by

$$N(L) = \sum_{m=1}^{N}\left(\frac{M^2}{m}\right)p(m,L) = M^2 \sum_{m=1}^{N}\frac{1}{m}p(m,L)$$

As $N(L) \propto L^{-D}$

Hence $\displaystyle\sum_{m=1}^{N}\frac{1}{m}p(m,L) \propto L^{-D}$

Let $N'(L) = \displaystyle\sum_{m=1}^{N}\frac{1}{m}p(m,L)$

Hence $N'(L) \propto L^{-D}$

The fractal dimension D can be calculated by finding the slope of a log-log curve for a series of values of N'(L) and L. To evaluate P(m,L), a box of side L is moved around each point within the box. For every point find out the value. Find out the frequency of occurrence of each m and divide it by the total number of points N within the box. This gives the P (m,L) value of that particular m.

7.3.3 Differential Box-Counting Method

Sarkar et. al. [94] gave another method to find out the fractal dimension known as Differential Box Counting (DBC) method. Instead of counting the boxes like the cell counting method, they found out the minimum & maximum gray level of the image in the (i, j)th grid, which may fall in the k and l box respectively. Then $n_r(i, j) = l-k+1$ is the contribution of the N_r in the (i, j)th grid. Taking contributions from all the grids $\mathbf{N_r}$,

$$N_r = \sum_{i,j}n_r(i,j)$$

Methods Images ➡	Cell Counting Method	Keller's Method	Differential Box Counting
Overlapping Triangles	2.167198	2.441961	2.583887
Pyramid Furnace	2.211895	2.461466	2.676281
Open Wheel	2.176739	2.669181	2.578142
Twin Kite	2.12107	2.81634	2.382044
Hanging Tower	2.108768	2.487009	2.343462

Table 7.2: Fractal dimensions of images obtained by various methods

N_r is counted for different value of r (i.,e. in turn different values of L). Then D can be estimated from the least square linear fit of $\log(N_r)$ against $\log(1/r)$.

Bisoi and Mishra made a comparison between the various box-counting dimensions [4] which are shown in table 7.2. For this purpose the images generated by the authors were considered [5].

7.4 Bound of the box size

Box counting as we know is a very simple process. However, many researches have been performed to improve the procedures with respect to the calculation of the roughness accurately. Some times problems occur while experimenting the procedure effectively due to improper limits and box size. Of course, corrective measures have been taken by the authors. Many authors have assumed certain bounds according to the procedure adopted by them. Sarkar et. al. [20,94] took a bound as $2 \le L \le M / 2$ in his differential box counting method.

7.4.1 Keller's Correction

Keller et. al. worked on the same aspect and also established a theoretical justification for the lower limit for the box size [120]. They established it by stating that as the box size decreases, the number of boxes needed to cover the fractal set will increase according to the power law. However the maximum number of nonempty boxes will be equal to the total number of discrete points. Thus if L_B is the lower bound of the box size, then by power law:

$L_B \propto (1/M)^{1/D}$

where M is the total number of points on the image and D is the fractal dimension. Further from the self-similarity property of fractals, if the entire image is contained within one box of size L_{max}, then the image S is scaled down by a ratio r will have the property

$Mr^D = 1$

i.e $M = (1 / r)^D$ for box of size L.

Also $L = rL_{max}$

Hence $N(L) = (1/r)^D = (L_{max}/L)^D$ is the generalised formula.

By arithmatic manipulation, we have

$$L^D = \frac{\left(L_{max}\right)^D}{N(L)}$$

$$i.e \; L = \left[\frac{(L_{max})^D}{N(L)} \right]^{1/D} = \frac{(L_{max})^D}{[N(L)]^{1/D}}$$

When the box size L goes on decreasing to a lower bound **L$_B$**, N(L) will be equal M, then

$$L_B = \frac{L_{max}}{(M)^{1/D}}$$

Keller et. al. also proposed a procedure to first find out D with different L values and then follow an iterative procedure to produce L$_B$ value using the above equation and re-estimate the D for only box sizes of L > L$_B$. This iteration should continue till D is found out based on the box size of L$_B$. Of course, he stated that the process will converge for ideal fractals only.

7.4.2 Fractional Box-Counting Method

Lin et. al. [65] argued over the counting process. They pointed out that under counting will occur in small scales and over-counting will occur in large scales in case there are uniform intensity areas. Because though the intensity variance is zero for such areas and hence their contribution towards the fractal dimension is zero, but they do have certain value. Hence incase of small scales, such areas are ignored & and this results in undercounting. They particularly emphasized on this. Similarly in case of large scales we may take into account uniform intensity areas also giving rise to over-counting. Therefore in order to avoid such problems, they proposed a fractional box counting method where the shape of the boxes can be deformed, depending on the intensity labels.

Let $A \in S(X^m)$. If $\rho(n) = 2^{-n}$ is the counting scale, define $\rho_0 = 2^{-n-q}$ as the base scale for n=1,2,3,..... & q=1,2,3,....... Cover the space X^m by closed adjacent square boxes of side $\rho(n)$. Break each box at counting scale $\rho(n)$ further into boxes at base scale ρ_0. Let $\eta_n(A)$ and $\eta_0(A)$ denote the number of boxes of size $\rho(n)$ and ρ_0 which intersect the set X^m respectively. If A has box counting dimension D then

$$D' = \lim_{n \to \infty} \{ \ln(\eta_0[A]) / \ln(1/\rho_0) \} = D$$

exists, and **D'** is called the fractal box dimension of A.

7.4.3 Mishra's Bounds

Bisoi and Mishra [4] considered the fundamental aspect of roughness, which implies a fractal dimension of 3 for the roughest possible surface. Always, $L \times L = L^2$ pixels on a particular grid corresponds into all M / L number of vertical boxes on that grid. In order to obtain maximum roughness all the boxes on top of each grid should be countable, as in this case the total number of countable boxes are

N = No. of grids $\times$ M / L

=M / L $\times$ M / L $\times$ M / L

=[M /L]^3

Then the dimension D = 3 as 1 / r = M / L.

However, sometimes it may so happen that $L^2 < M / L$ and all M / L boxes cannot contain at least one gray level intensity surface. Hence in this case the possibility of getting maximum roughness has to be ruled out. Therefore, in order to obtain maximum roughness, the box size **L** should be considered such that

$L^2 \geq M / L$

$\Rightarrow L^3 \geq 2M$

Of course, roughness of the image is obtained by considering various L values and drawing a log-log plot of N versus L. But still, the lower L values, which don't obey the above equation will result in some incorrect points on the log-log plot and hence an incorrect D value. Therefore, the different L values should be taken such that the above equation is satisfied.

Regarding the upper limit of the box size, the maximum box size should be up to M / 2. Otherwise, even if L value is increased more than M / 2. only one grid on the entire image will cover a part of the image and hence N cannot be counted. Thus, L should be chosen such that

$L \leq M / 2$

Therefore, the value of L should be chosen such that it satisfies both the above equations, which will provide the accurate roughness of any image.

7.5 Multifractals

So far the studied objects are fully characterised by a single value D, the Hausdorff dimension, hence forth denoted by D_0. Fractals found in nature however, most often need for their characterisation a whole spectrum of dimensions called generalised dimensions Dq. The Hausdorff dimension is only one dimension in this continuous spectrum. Such fractals are called multifractals. The essential difference between the construction of multifractals and one-scaled fractals is that the initial set is divided into N not identically sized parts. Each sub-set is a reduced version of the original object by some factor.

7.5.1 Multifractal Point Patterns

Kenkel considered a spatial pattern of N points. A grid of boxes of length ϵ is laid over the pattern and a count of the number of points in each of the N_ϵ occupied grid boxes is determined as shown in table 7.3 and 7.4 below. Then each box count p_i can be expressed as a proportion:

$P_i = n_i / N$

where n_i is the number of points in the ith grid and N is the total number of points.

The generalized entropy is defined as

$$I_q(\epsilon) = 1/(1-q) \log \sum_{i=1}^{N_\epsilon} p_i^q$$

By varying q, an entire family of entropy functions is defined as given in table 7.5. The generalized dimension $\mathbf{D_q}$ for the qth fractal moment is given by

$$\mathbf{D_q} = - \lim_{\epsilon \to 0}\left[I_q (\epsilon / \log(\epsilon))\right]$$

2	26	32	1				
17	55	8		11	28		2
37	30		22	74	78	24	44
2	70	2	6	31	29	12	19

Table 7.3 Number of points per box of size 4

				4								
		2	6									
	3	8										
						0	9					
					1	5	6	6			3	3
4	0				0		1		6	2		
	0	0					6	5			8	
	6	4					5					

Table 7.4: Number of points per box of size 2

Q	$I_q(\in)$	Name	Dimension
0	$\text{Log } N_\in$	Log (box count)	Box-count
1	$-\Sigma p_i \log p_i$	Shannon entropy	Information
2	$-\log \Sigma p_i^2$	Log (Simpson index)	Correlation

Table 7.5: Entire family of entropy functions

D_q is determined from the slope of the $I_q(\in)$ vs. $\log \in$ plot. For a classic fractal object D_q is a simple linear function of q. For multifractal objects, the relationship between D_q and q is non-linear. For classic fractal, all moments give the same value of D. On the other hand, multifractals have a different value of D for each moment.

7.5.2 Multifractal Calculations on Generalized Sierpinski Triangles

Falconer et. al. [77] gave a theorem to find out the box-counting dimension of the generalized Sierpinski triangle is given by the blueprint in figure 7.5, where the IFS (S_1 is a similarity) is given as below.

The parameters are fixed as $0 < a$, $b < 1$. The box-counting dimension **s** can be found out from the following formula.

$(1-a)^2 + ab^{s-1} + a(1-b)^{s-1} = 1$

whereas a $\geq 1-b$ and a $\geq b$.

The generalized Sierpinski triangle set supports the self-affine multifractal measure μ such that for an IFS $\{S_1, S_2, \ldots\ldots S_k\}$ on R^n with associated probabilities $\{p_1, p_2, \ldots\ldots p_k\}, > 0$ and $\Sigma p_i = 1$,

$$\mu(B) = \sum_{i=1}^{k} P_i \mu(S_i^{-1}(B))$$

Multifractal analysis $\beta(q)$ of self-similar measures [18, 19, 20,22] shows that

$$\sum_{i=1}^{k} p_i^q r_i^{\beta(q)} = 1$$

Hence the generalized dimension can be given by

$$D_q = \frac{\beta(q)}{1-q}$$

For q ≠ 1.

$$S_1 \begin{bmatrix} x_1 \\ x_2 \end{bmatrix} = \begin{bmatrix} 1-a & 0 \\ 0 & 1-a \end{bmatrix} \begin{bmatrix} x_1 \\ x_2 \end{bmatrix} + \begin{bmatrix} 0 \\ a \end{bmatrix}$$

$$S_2 \begin{bmatrix} x_1 \\ x_2 \end{bmatrix} = \begin{bmatrix} b0 \\ 0a \end{bmatrix} \begin{bmatrix} x_1 \\ x_2 \end{bmatrix} + \begin{bmatrix} 0 \\ 0 \end{bmatrix}$$

$$S_3 \begin{bmatrix} x_1 \\ x_2 \end{bmatrix} = \begin{bmatrix} 1-b & 1-a-b \\ 0 & a \end{bmatrix} \begin{bmatrix} x_1 \\ x_2 \end{bmatrix} + \begin{bmatrix} b \\ 0 \end{bmatrix}$$

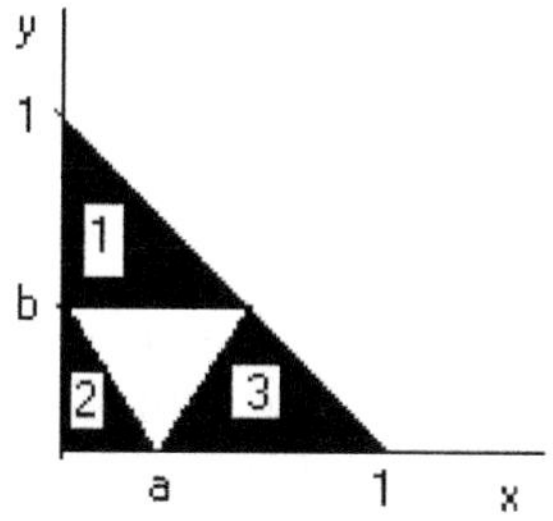

Figure 7.5: Blueprint for generalized Sierpinski triangle

7.6 Fractal dimension from the L-System

The fractal dimension can be calculated as a ratio between the fractal curve's growth in length and how much it advances [86]. Although this concept has not been tested with complex fractal curves from the equivalent L-System, a few fractals have been experimented with, which give same numerical value for dimension as calculated by using grid dimension method as discussed. So fractal dimension from the L-System has been calculated using the formula given below;

$$D = \frac{\log(N)}{\log(d)}$$

Where, N, is the length of visible walk that follows the fractal generator and d is the distance in a straight line from the start to the end point of the walk measured in turtle step lengths. The algorithm computes two numbers. The first is the length N of the visible walk that follows the fractal generator, which is equivalent to the number of graphic symbols, i.e., the number of symbols F. The second is the distance d in a straight line from the start to the endpoint of the walk, measured in turtle step length, which can also be calculated from the string.

Figure No	L-System Codes	Generator String	N	d	Dimension
	Axiom = F--F--F F=F+F--F+F	F+F--F+F	4	3	1.261

Table 7.6: Fractal dimension of Koch curve from L-System

From this above table, it is clear that fractal dimension of Koch curve as calculated using L-system codes is these two different methods are same which is in accord with the results obtained by other methods.

7.6.1 Fractal dimension of hybrid fractals from L-System

Fractal dimensions of the hybrid fractals as shown in the Table 4.15 of the Chapter-4 are given below in the Table 6.15.

From the table it is observed that the dimension of the fractals through L-System for a particular figure is almost same as that to be calculated by the traditional definition. This may be due to the fact that the length N of the visible walk may not be equal to the number of draw symbols in the generator string. So in this stage the fractal dimension from equivalent L-System may not be considered as a standard method for determining the fractal dimension.

However this method may be considered as the basis for determining the fractal dimension from L-System strings directly. In future this method can certainly be extended to calculate the complicated turtle graphics interpreted L-System strings.

It is clear that the hybrid fractals are unique in nature although they have been developed from the base classical fractal.

Fig. No	L-System Axiom	Generator String	N	d	Dimension
Fig.(a)	Axiom = A	A=AF--F--F|BF+F--F+F	7	5	1.20962
Fig.(b)	Axiom = A	A=AF-F-F|BF[-F]F	6	5	1.113283
Fig.(c)	Axiom=AF-F-F-F	A= AF-F-F-FBF-F+FF-F-F+F	11	10	1.041393
Fig.(d)	Axiom = A	A=AF--F--F|BF+F--F+F	7	5	1.20962
Fig.(e)	Axiom = A	A=AF-F-F|BF [-F] F	6	5	1.113283
Fig.(f)	Axiom = A	A=AF-F-F-F|BF-F+F+FF-F-F+F	12	11	1.036289
Fig.(g)	Axiom = A	A=AF--F--F|BF+F--F+F	7	5	1.20962
Fig.(h)	Axiom = A	A=AF-F-F-F|BF-F+F+FF-F-F+F	12	11	1.036289
Fig.(i)	Axiom=AF-F-F-F	A= AF-F-F-FBF-F+FF-F-F+F	11	10	1.041393
Fig.(j)	Axiom = A	A=AF--F--F|BF+F--F+F	7	5	1.20962
Fig.(k)	Axiom = A	A=AF--F--F|BF+F--F+F	7	5	1.20962
Fig.(l)	Axiom= AF-F-F-F	A= AF-F-F-F BF-F+FF-F-F+F	11	10	1.041393

Table 7.7: Fractal dimensions of hybrid fractals from L-System

7.7 Summary

This chapter gives an insight into the estimation of fractal dimension of various fractals. Starting from the Mandelbrot's formula to the box counting dimension it provides a discussion of the fractal dimension estimation. Finally it gives an estimation of fractal dimension through L-system.

Chapter 8

Research Directions of L-Systems

L-systems have been used in many scientific areas. Selected applications with ongoing research work and future directions for research progress have been highlighted in this chapter so as to make all readers and researchers perceive the application of L-System concept in their own research. Using the L-System approach one can simulate almost all fractal graphics concepts. This shows the importance of the L-System concept. In this chapter large number of research works have been mentioned with the author and other related works. The bold heading in subsequent subsections of sections indicates the title of the paper published by the author. The detailed information regarding the journals or conferences where these papers have been published can be searched from the internet resources by the author's name or the title of these papers.

8.1   L-Systems on Data Base Management Systems

One of the most important advantages of database systems [36,134] is that the underlying mathematics is rich enough to specify very complex operations with a small number of statements in the database language. This research covers an aspect of biological informatics that is the marriage of information technology and biology, involving the study of real-world phenomena using virtual plants derived from L-systems simulation. Aristid Lindenmayer as a mathematical model of multicellular organisms introduced L-systems. Not much consideration has been given to the problem of persistent storage for these simulations. Current procedures for querying data generated by L-systems for scientific experiments, simulations and measurements are also inadequate. To address these problems the research in this paper presents a generic process for data-modeling tools (L-DBM) between L-systems and database systems. Authors [134] showed how L-system productions can be generically and automatically represented in database schemas and how a database can be populated from the L-system strings. The research further describes the idea of pre-computing recursive structures in the data into derived attributes using compiler generation. A method to allow a correspondence between biologists' terms and compiler-generated terms in a biologist-computing environment is supplied. Once the L-

DBM gets any specific L-systems productions and its declarations, it can generate the specific schema for both simple correspondence terminology and also complex recursive structure data attributes and relationships.

Another method [135] has been designed by using generic Data Model Tools for L-Systems (DML). The work combined a novel application area with databases in Bioinformatics. DML allows specification of database structures and queries in terms of objects and relations specific to scientific L-system applications. L-systems have been used in many scientific areas. However, not much consideration has been given to persistent storage and querying of the large quantities of data resulting from these L-system simulations.

Selected categories have been made with abstracts in subsequent sections for diversified thinking of a researcher.

8.1.1   Retrieval, Indexing & storage

Fractals for Secondary Key Retrieval

Christos Faloutsos and Shari Roseman proposed the use of fractals and especially the Hilbert curve, in order to design good distance-preserving mappings. Such mappings improve the performance of secondary-key- and spatial-access methods, where multi-dimensional points have to be stored on a 1-dimensional medium (e.g., disk). With L-System approach a researcher can reduce the access time.

Hilbert R-tree: An Improved R-tree Using Fractals

Christos Faloutsos and Ibrahim Kamel proposed a new Rtree structure that outperforms all the older ones. The heart of the idea is to facilitate the deferred splitting approach in R-trees. Proposing an ordering on the R-tree nodes does this.

FIRE: fractal indexing with robust extensions for image databases

Distasi, R. Nappi, and M. Tucci, M. presented FIRE, an image-indexing algorithm. As already documented in the literature, fractal image encoding is a family of techniques that achieves a good compromise between compression and perceived quality by exploiting the self-similarities present in an image. Because of its compactness and stability, the fractal approach can be used to produce a unique signature, thus obtaining a practical image indexing system for use with large databases.

8.1.2 Queries

The Fractal Dimension: Making Similarity Queries More Efficient

Adriano S. Arantes, Marcos R. Vieira, Agma J. M. Traina, and Caetano Traina presented a new algorithm to answer k-nearest neighbor queries called the Fractal k-Nearest Neighbor (k-NNF ()). The algorithm estimate a suitable radius to shrinks a query that retrieves the k-nearest neighbors of a query object. Experiments performed with real and synthetic datasets over the access method Slim-tree show that the total processing time drops up to 50%, while requires 25% less distance calculations.

Deflating the Dimensionality Curse Using Multiple Fractal Dimensions

Bernd-Uwe Pagel, Flip Korn and Christos Faloutsos discussed that nearest neighbor queries are important in many settings, including spatial databases (Find the closest cities) and multimedia databases (Find the most similar images). Previous analyses have concluded that nearest neighbor search is hopeless in high dimensions, due to the notorious "curse of dimensionality". They show real data exhibit intrinsic ("fractal") dimensionalities that are much lower than their embedding dimension and were able to 'deflate' the dimensionality curse via fractal analysis.

Estimating the Selectivity of Spatial Queries Using the Correlation Fractal Dimension

Alberto Belussi and Christos Faloutsos examined the estimation of selectivities for range and spatial join queries in real spatial databases and predict the selectivity of spatial joins using the Correlation Fractal Dimension. It is shown that real point sets: (a) violate consistently the "uniformity" and "independence" assumptions, (b) can often be described as "fractals", with non-integer (fractal) dimension.

Beyond Uniformity and Independence: Analysis of R-trees Using the Concept of Fractal Dimension

C. Faloutsos and I. Kamel proposed the fractal dimension of a set of points to quantify the deviation from the uniformity distribution, showing that real data behave as mathematical fractals, with a measurable, non-integer fractal dimension. The authors provide the first analysis of R-trees for skewed distributions of points, develop a formula that estimates the number of disk accesses for range queries, given only the fractal dimension of the point set, and its count. It suggested that the fractal dimension would help replace the uniformity and independence assumptions, allowing more accurate analysis for any spatial access method, as well as better estimates for query optimization on multi-attribute queries.

8.1.3 Text Summarization

Fractal Summarization: Summarization Based on Fractal Theory

Christopher Yang and Fu Lee Wang introduced a text summarization model based on the hierarchical structure of documents. Information is extracted from both document structures and on-page factors. The end result is a reduced copy of the original document. The authors found that fractal summarization outperforms traditional summarization algorithms found in the IR literature.

Fractal Summarization for Mobile Devices to Access Large Documents on the Web

Christopher Yang and Fu Lee Wang introduced the fractal summarization model for document summarization on handheld devices. It generates a brief skeleton of summary at the first stage, and the details of the summary on different levels of the document are generated on demands of users. Such interactive summarization reduces the computation load in comparing with the generation of the entire summary in one batch by the traditional automatic summarization, which is ideal for wireless access.

8.1.4 Language, Semantics and Relevancy

Neuro-Fractal Composition of Meaning: Toward a Collage Theorem for Language

Simon Levy presents a formal mathematical model for putting together words and phrases, based on the iterated function system method used in fractal image compression. Building on vector-space representations of word meaning from contemporary cognitive science research, the author shows how the meaning of phrases and sentences can likewise be represented as points in a vector space of arbitrary dimension.

Valuations of Languages, With Applications to Fractal Geometry.

Henning Fernau used Iterated Function Systems (IFS) in a valuation model. The paper shows that valuations are useful not only within the theory of codes, but also when dealing with ambiguity, especially in context-free grammars, or for defining outer measures on the space of w-words which are of some importance to the theory of fractals.

The fractal nature of relevance: a hypothesis

Jim Ottaviani proposed a fractal geometry model for clusters of relevant documents. It reflects the relatively simple iterative search process used by interactive online searchers. Clusters formed using dynamic search strategies appear topologically distinct, indecomposable, and result from chaotic processes.

Recognition and Generation of Fractal Patterns by using Syntactic Techniques

Jacques Blanc-Talon showed the connection between D0L-systems (a special type of free-grammar L-System), language and fractals. The problem of inferring a Context-Free Grammar containing DOL-like and free recursive structures is discussed.

Pavements as Embodiments of Meaning for a Fractal Mind

Terry M. Mikiten, Nikos A. Salingaros and Hing-Sing Yu have written a paper that puts forward a fractal theory of the human mind that explains one aspect of how we interact with our environment. Analogies have been developed for storing ideas and information within a fractal scheme. The mind establishes a connection with the environment by processing information, this being an important theme seen during the evolution of the brain.

Universal Grammar

Charles Henry discussed the fractal nature of grammar as it relates to word associations and cognitive associative patterns.

8.1.5 Latent Semantics & Reduction Techniques

Fractal Dimension for Data Mining

Krishna Kumaraswamy applied applications of fractals to Latent Semantic Indexing and the problem of dimensionality reduction. He introduces the concept of intrinsic fractal dimension of a data set and shows how this can be used to aid in several data mining tasks.

A Study in Fractal Dimension and Dimensionality Reduction

Elena Eneva, Krishna Kumaraswamy and Matteo Matteucci investigated the relationship between several dimensionality reduction methods and the intrinsic dimensionality of the data in the reduced space, as estimated by the fractal dimension.

Optimal Fractal Coding is NP-Hard

Matthias Ruhl and Hannes Hartenstein demonstrated by a reduction from MAXCUT that the problem of determining the optimal fractal code is NP-hard. In fractal compression (reduction) a signal is encoded by the parameters of a contractive transformation whose fixed point (attractor) is an approximation of the original data. Fractal coding can be viewed as the optimization problem of finding in a set of admissible contractive transformations the transformation whose attractor is closest to a given signal.

8.1.6 Datamining

Fractal Databases - New Horizons in Database Management

Lisa Lewinson's online version of an article published in PC AI (March/April 1994). In this article she presented many benefits of designing and using fractal databases (FD). Specific applications to business FD's are discussed.

Using the Fractal Dimension to Cluster Datasets

Daniel Barbara and Ping Chen presented a new cluster-ing algorithm, Fractal Clustering (FC), which places points incrementally in a cluster for which the change in the fractal dimension after adding the point is the least. FC requires one scan of the data, is suspendable at will, providing the best answer possible at that point, and is incremental. FC effectively deals with large data sets, high-dimensionality and noise and is capable of recognizing clusters of arbitrary shape.

Tracking Clusters in Evolving Data Sets

Daniel Barbara and Ping Chen presented an algorithm to track the evolution of cluster models in a stream of data. The algorithm is based on the application of bounds derived using Chernoff's inequality and makes use of the Fractal Clustering algorithm (FC), which uses self-similarity as the property to group points together. Experiments show the tracking algorithm is efficient and effective in finding changes on the patterns.

Geospatial Databases and Data Mining

Report of several agencies discussing how fractal pattern techniques can be used to deal with spatio-temporal data, geo-data, data structures, queries, indexes, and algorithms.

Fractals and Self-similarity in Data Mining: Issues and Approaches

8th ACM SIGKDD International Conference on Knowledge Discovery and Data Mining dedicated to data mining techniques through fractal dimensions and self-similar characteristics in different domains. Topics discussed: dimension reduction, predictive modeling, using self-similar characteristics to mine databases such as association rules, clustering, and classification, modeling and finding outliers, selectivity estimation, spatial databases, R-trees, Quadtrees and model distributions of data.

How to Use the Fractal Dimension to Find Correlations between Attributes

Elaine Parros Machado de Sousa, Caetano Traina Jr., Agma J. M. Traina and Christos Faloutsos presented an algorithm to select the most important attributes (dimensions) for a given set of n-dimensional vectors, determining what attributes are correlated with the others and how to group them. The algorithm uses the 'fractal' dimension of a data set as a good approximation of its intrinsic dimension and, based on it, indicates what attributes are the most important.

Accelerating Clustering methods through Fractal based Analysis

Changhao Jiang, Yiheng Li, Minglong Shao and Peng Jia presented a fractal analysis procedure for accelerating clustering methods by sampling dataset into critical-sized subset with preserve the original set distribution patterns. Experiments with the BIRCH clustering method are also discussed.

Self-Similar Layered Hidden Markov Model

Jafar Adibi, Wei-Min Shen presented SSLHMM introduce, analyze and demonstrate Self-Similar Layered HMM (SSLHMM), for a certain group of complex problems which show self-similar property, and exploit this property to reduce the complexity of model construction. SSLHMM reduces the complexity of learning and increase the accuracy of the learned model.

Time-Invariant Sequential Association Rules: Discovering Interesting Rules in Critical Care Databases

Jafar Adibi and Wei-Min Shen provided formalism for the discovery of self-similar association rules in critical care databases.

8.2 Parsers, Compilers

Dynamical Parsing to Fractal Representations

Simon Levy presents a connectionist-parsing model in which traditional formal computing mechanisms (Finite-State Automaton; Parse Tree) have direct recurrent neural-network analogues (Sequential Cascaded Net; Fractal RAAM Decoder). The model is demonstrated on a paradigmatic formal context-free language and an arithmetic-expression parsing task.

Fractal Symbolic Analysis

Vijay Menon, Keshav Pingali and Nikolay Mateev proposed a new form of symbolic analysis and comparison of programs for use in restructuring compilers. Fractal symbolic analysis is an approximate symbolic analysis that compares a program and its transformed version by repeatedly simplifying these programs until symbolic analysis becomes tractable while ensuring that equality of the simplified programs is sufficient to guarantee equality of the original programs.

8.3 Web Links

The Fractal nature of the Web

Tim Berners-Lee discovered that the structure of the Web is fractal.

Self-Similarity in the Web

Kumar, Dill, Mccurley, Rajagopalan, Sivakumar and Tomkins showed that the Web emerges as the outcome of a number of essentially independent stochastic processes that evolve at various scales. A striking consequence of this scale invariance is that the structure of the Web is fractal. An understanding of this underlying fractal nature is therefore applicable to designing data services across multiple domains and scales.

Are we able to characterize Semantic Web behaviour?

Rosa Gil, Roberto Garcia and Jaime Delgado found self-similarity and power law behaviors in Self-Organized Critically, SOC, Complex Systems, CS Patterns, Barabasi's Scale-Free Networks (BA Model) and the Semantic Web.

Fractal-small-world dichotomy in real-world networks

Gabor Csanyi and Balazs Szendroi presented the fractal-small-word dichotomy that exists between small-world networks exhibiting exponential neighborhood growth, and fractal-like networks, where neighborhoods grow according to a power law. The distinction is observed in a number of real-world networks, and is related to the degree correlations and geographical constraints. The authors present a simple method for examining scaling behaviors of small-connected graphs. Applications to internet traffic, the Web and industrial networks are possible.

Online Generation of Fractal and Multifractal Traffic

Darryl Veitch, Jon-Aders, Jens Wall, Jennifer Yates and Matthew Roughan used a general wavelet framework to describe the online generation of time series, particularly fractal and multifractal time series. A scalable system is presented to test internet traffic.

A General Fractal Model of Internet Traffic

Sandor Molnar and Gyorgy Terdik presented a fractal model of internet traffic. The fractal nature of Internet traffic has been observed by several measurements and statistical studies. In this paper a new monofractal stochastic process called Limit of the Integrated Superposition of Diffusion processes is presented.

Neural Network Modeling of Self-Similar Teletraffic Patterns

Homayoun Yousefi'zadeh described that the significant characteristic of bursty traffic is self-similarity. Self-similarity is the main reason for observing so-called burst within burst patterns across a wide range of time scales as one of the unique characteristics of nonlinear systems with fractal nature.

8.4 Tools

Visualization Tools for Self-Organizing Maps

Christopher C. Yang, Hsinchun Chen and K. K. Hong presented FISHEYE and FRACTALVIEW. Self-organizing category map is identified as a powerful tool for information summarization. However, visualizing a large-scale self-organizing map in a restricted size of window is difficult. For smaller regions, displaying labels is infeasible. The tools assist users to visualize a large-scale self-organizing map geographically and semantically.

SELFIS: A Tool For Self-Similarity and Long-Range Dependence Analysis

Thomas Karagiannis and Michalis Faloutsos presented a java-based tool for use in fractals, self-similarity, long-range dependence, power-laws and Hurst Exponent studies. Evidence of fractals, self-similarity and long-range dependence in network traffic is discussed.

Wildwood: The Evolution of L-System Plants for Virtual Environments

This paper describes the Wildwood project. In this work, a genetic algorithm was applied to a simplified L-system representation in order to generate artificial-life style plants for virtual worlds. Acting as a virtual gardener, a human selects which plants to breed, producing a unique new generation of plants. An experiment involving a simulation-style fitness function was also performed, and the virtual plants adapted to maximize the fitness function.

Due to the recent growth of the Internet, personal computers have been transformed into communications devices. This is exemplified by the growth of virtual chat spaces that began with text chat and have moved to 2d and 3d graphical environments that now incorporate a large degree of realism. Users worldwide populate these virtual worlds in the form of avatars. Most of these worlds are very antiseptic and contain only static manmade objects.

Wildwood represents exploratory technology created with the intent to enrich virtual worlds. Its current incarnation is akin to genetic evolutionary art. The genetic operations allow a virtual gardener to explore new variations of plants. The system also allows novices to play with and evolve plants that suit their preferences without the need to learn about L-systems or production rules.

Many enhancements are possible that can generate more interesting and varied plants. Foremost is the addition of new terms to control segment width and color. Moreover, the incorporation of multiple production rules would also support the generation of new classes of plants. Individual rules may be trained independently to form a hierarchical system. For example, one set of rules may form leaves while another may form flowers or roots.

Another enhancement is the addition of parameters to the genetic operators. Currently, the GA only operates on production rule strings. Variable rotation angles, starting angles, and order could produce other interesting virtual plants at different stages of development if they were parameters that the GA could adjust.

As a simulation tool, a great deal of work needs to be done modeling the environment and the effect that the environment has on plants in order to produce useful results. In the near term, this is most easily accomplished by focusing on specific environmental factors, such as wind, rain, or sunlight.

Overall, many exciting opportunities remain to be explored from both an aesthetic and scientific perspective. Wildwood is a first step towards uncovering new possibilities and examining the resulting issues.

8.5 Recent field of Applications of L-Systems

This section has documented few challenging and less investigated topics in the field of L-system.

8.5.1 Indexing and Data Mining in Multimedia Databases

Professor Christos Faloutsos (Carnegie Mellon University) highlighted many limitations of traditional information retrieval methods and introduces new tools for datamining: Fractals. Discusses also FastMap (for visualization) and Falcon (for relevance feedback). Includes Internet topology, applications to sales and financial data as well as merging of similarity scores from dissimilar multimedia objects. He discusses how the "dimensionality reduction course" of SVD (LSI) can be avoided with fractal techniques.

8.5.2 L-Systems and Fractal Growth

Structures in natural systems (plant growth, animal circulatory systems, etc.) often reflect fractal geometries. These fractals in biological systems must be grown, and the "programs" are found in DNA. One can investigate some of the formalisms and programs for the fractals. The fun part is seeing what the fractals produce. Note that this does not require prior knowledge of any "programming language" - anyone can learn this from scratch. Most of the research can be done using the Web and the library.

8.5.3 Cellular Automata

A cellular automaton (CA) is a type of computing machine invented by John von Neumann, one of the fathers of computer science. A CA is defined by a grid (it could be one-dimensional, or two, or more), with each element of the grid being "on" or "off", and a set of rules that describe how each element changes based on the states of neighboring elements. Despite this simple description, CAs is capable of a wide range of behavior and capable of universal computation (i.e., as "powerful" as any computer). In this area, one can explore the various types of CAs and their behaviors. This area does not require prior knowledge of any programming language - anyone can learn this from scratch. Most of the research can be done using the Web and the library.

8.5.4 Autonomous Agents

An autonomous agent (AA) is a simple entity that interacts with its environment and other AAs, typically based on a simple set of rules. For example, the AAs may be birds that randomly fly around a grid, but obey simple rules like avoiding bumping into each other and not flying directly behind another bird (otherwise it cannot see). It is interesting to then observe the "emergent behavior," such as the patterns of movement. Researchers have developed very simple rules that seem to mimic the patterns seen in nature of how birds fly in formations. In this area, one can explore the various types of AAs and their behaviors. This area does not require prior knowledge of any programming language - anyone can learn this from scratch. Most of the research can be done using the Web and the library.

8.5.5 Explore Individual Differences in Handwriting

Handwriting experts can look at a sample piece of handwriting and determine whether it is authentic or a forgery. They accomplish this task by looking at the way the words are written, i.e. the thickness of the stroke, the slant of the letters, the spacing between the letters, etc. Each of these these characteristics are called a *feature* of the handwriting. One can use L-systems approach to solve this problem.

8.6 Classifiers, IFS and L-Systems

Classifiers, Iterated Function Systems and Lindenmeyer Systems form part of a range of complexity techniques known as Production Systems. One can observe these ideas and see how they can be used to allow systems to develop over time in a contextual way. A production system is a set of rules that trigger each other to produce action in their environment and as such could be seen as limited models of how cells, organisms or brains work. The general nature of these models allows them to be applied in many different forms.

8.6.1 Iterated Function Systems (IFS)

An iterated system departs from the linear input $\rightarrow$ process $\rightarrow$ output chain common to most scientific thinking and loops the output back into the input. Thus the same process occurs repeatedly (recursively). As the process can modify the system variables this allows the results to cover a wide range or to map out a path or orbit in state space. Fractal equations like the Mandelbrot and Julia sets are systems of this type. In IFS Fractals the system consists generally of a set of mathematical equations (usually contractive affine transformations) one of which is chosen (usually probabilistically) for each iteration. This process generates such famous constructions as the snowflake, Sierpinski triangle and Barnsley fern.

We can say that in general IFSs are the attractors that emerge when assembled from small copies of themselves, using combination of translations, rotations, and scalings. They exhibit strong self-similarity. It is the 'choice' between transformations aspect of IFS that distinguishes them from normal linked equations (although conditionals can also be inserted into fractal equation evaluation - e.g. Fractint 'formula'). IFS systems are used extensively in data compression applications (the IFS rules form a 'seed' that concisely encodes a picture).

8.6.2 Lindenmeyer Systems (L-Systems)

To make the system more flexible IF... THEN decisions can be added and in this form it approaches that of expert systems where only part of the 'rule book' can be implemented at any one-time step. L-Systems thus add context to the IFS production scheme. Here we have a list of transforms between symbolic components. For example we may say component A goes to B (IF A THEN B), and B goes to AA. Applying these rules in turn to an initial seed (say A), we get the growing sequence:

A

B

AA

BB

AAAA

BBBB

AAAAAAAA

These are formal context-sensitive grammars consisting of Variables, Constants, Rules and Initialisations, which specify a set of rewriting rules. More complex versions of these rules are often designed to incorporate geometric elements (Turtle graphics or LOGO), which can allow decisions such as "IF there is an A to the left AND a B to the right MOVE C steps forward ELSE ROTATE D degrees" to generate pictures. These are further improved if random noise (or environmental sensitivity) is added to avoid mechanical regularity. This process can give rise to amazingly realistic creations of natural organic forms as seen in this example of a plant. These forms of directed building block recombination provide methods of growth or development, ways of exploring state space, the space of possible structures.

8.6.3 Schemas

Both IFS and L-Systems are static sets of rules in essence, but we can look to evolving them by the techniques used in Genetic Algorithms. GAs generally use a two state alphabet of 1 and 0 to specify genotype components, but we can generalise this to include also a 'don't know' character (#), By doing this we can have a ternary (3 state) string that specifies more than one possible genome or rule. For example for a 4 bit string, #### specifies all of state space, 1### all genomes beginning with a 1. Note that a single specified character partitions the state space in half (it will match half the possible genomes), this is the order or specificity of the rule, thus a rule of form #10# has order 2 and 1011 has order 4. Each rule is called a schemata and the set of rules (partitions over state space) the schema.

The most important aspect of schemas is Holland's 'Schema Theorem' which states: "Schemas whose average fitness remains above the population average will receive exponentially increasing numbers of samples over time". This means that the production system will operate in an implicitly parallel way, searching state space, and will converge to an optimum state (but not necessarily the global optimum). This parallelism is a form of weak cooperation, in that bit positions that work well together can be said to cooperate by association, so this methodology incorporates epistatic or synergistic associations between the bits (i.e. it takes into account trade-offs or compromise between the components).

8.6.4 Classifier Systems (CS)

Classifiers add the concept of schemas to the production rules, this enables us to generalise rules over the environment space and arrange to mutate or discover new rules with which to explore this space. These systems are mid way between non-symbolic systems such as Neural Networks and symbol-based systems such as Expert Systems and L-Systems. Central to the principle is the idea of a 'broadcast language' where each matched rule issues a general message (its output) to the environment (similar to how a gene 'broadcasts' a protein in a cell).

In these systems the environment is assumed to be sensed and each sensor codes a bit in environment space - rule space is thus mapped one-to-one onto this input space. While just creating different rule combinations allows us to explore rule space, it does not however tell us anything about the efficiency of the rules or the messages being used (they are thus a form of random collision model). In the GA world an additional layer is added to evaluate the fitness of each choice and to determine which genomes to discard and which to modify. For a practical classifier system some procedure of this type (called 'credit assignment or allocation') will also be necessary.

8.6.5 Reinforcement Learning (RL)

Various methods to evaluate rule schemes have been tried but, unlike GAs, which inhabits their own world, production systems relate to an external environment and thus any feedback regarding effectiveness must come from that source. Reinforcement Learning is the field that studies these ideas and indirectly includes both classifier systems and neural networks.

Two general forms of feedback are possible. In the first, the environment will give the 'correct' answer (rather like supervised learning in NNs or teachers), thus changes can be made directly to the system to better approximate the answer. Generally however the environment is not that specific and an answer simply of good or bad (or even of no response at all unless wrong) is more likely.

8.6.6   LCS Learning Classifier Systems

The general evaluation addition needed to form a simple classifier system (SCS) is the 'bucket brigade'. In this each rule has strength and if it matches an environmental input (e.g. input 1011, rule 1### matches, rule 0#10 does not) it takes its action. If this results in a correct response from the environment then the rule strength is increased. Generally there are many rules that match any input, so they all post their proposed responses to a 'message board' that also contains the environmental messages. The winning rule is that with the highest **strength time's specificity** (so that a more specific rule takes precedence over vague ones) with stochastic tie breaks. The best rules are mutated in some way to form new rules allowing the system to evolve a good set of rules. An important feature of classifiers is that they are strongly cooperative (they form chains of rules that are mutually fitness affecting and beneficial if successful - the 'bucket brigade')

Two main 'schools' of classifier thought are evident. In the Michigan approach each rule is treated as independent, which allows fast evolution but gives poor problem solving ability ('parasite' rules invade the population - a problem also seen in social individualism!). The Pitt approach treats the set of rules as a whole, having populations of different sets of rules which then compete. The winner here creates a more powerful set of cooperative rules but (like many collectives) is slow to find solutions and wasteful of resources. In these systems 'bid costs' and 'existence costs' can be used to penalise rules that are too general (always matched) or useless (wrong actions), which are eventually are then replaced.

8.6.7 Extended Classifier Systems

Due to the polarised nature of these two approaches alternatives have been proposed. One, Organized Classifier Systems (OCS) is based on the concept of transaction costs from economic theory. Here 'individuals' have high transaction costs (due to mistrust, duplication of work) but also flexibility, whilst 'collectives' have low costs (contracts) but suffer constraints. The OCS tries to balance these dynamic/static elements to gain a compromise ('edge of chaos') of optimum balance. We can regard the compromise 'organizations' here as variable length chromosomes, so a population of modular rule systems is possible.

Another approach is the XCS system, where strength is replaced by accuracy, allowing optimized populations of rules to arise in certain circumstances and by using a 'niched multiobjective genetic algorithm' these are applicable to multiple payoff landscapes also. This idea also allows planning to be incorporated (animate behaviour, applicable to situated robotics).

8.6.8 Finite State Machines (FSM)

The schemata considered up to now did not contain any explicit memory (the strength was an implicit memory but not used in the matching process), each message was interpreted in isolation from any previous system states. We can however generalise the idea of schemata and this brings us to the Finite State Machine. Like the classifier this responds to input messages and generates an output. The rule or 'transition table' specifies for each input combination what the output state should be, and this can be described as a logical function (normally using bivalent logic, but we can also have Fuzzy State Machines that use fuzzy logic). The inputs here can also be generalised to include 'tags' (identification strings) allowing machines (or 'agents') to recognise each other and perform type specific interactions.

One can incorporate memory into these multistate machines and if we do so then the output depends not only on the current input but on an history, which may be shallow (just the previous input) or rich (a whole life experience). In this form of decision unit we approach the Complex Adaptive System and ALife types of scenario. Despite the complexity of these components however we still usually assume a fixed population with an external method (the GA) for creating or deleting members.

8.6.9 Constrained Generating Procedures (CGP)

CGPs are a new formulation (also by Holland) of classifier systems, intended to generalise to modeling all forms of systems. In this formulation, rules (or procedures) are allowed to generate (or produce) new states within a constrained (limited scope) environment. Thus CGPs can be viewed as methods to compress global state space, ways of coding rules that can create all the states needed in any particular representation (e.g. a chess game or a Cellular Automata).

Classifier outputs cannot normally create new classifiers that are the role of the GA. But by relaxing that restriction we can arrive at a more general form of broadcast system, which can create its own parts. In these procedures the rules can specify not only actions on the environment but also actions on the agent structure itself. This variable form of constrained generating procedure (cgp-v) is potentially a universal constructor as well as a universal computer, since no external control or 'boss' component (like the GA) need now exist. Such an autonomous, free form and powerful agent is, needless to say, well beyond our current practical capabilities!

Whilst the simpler forms of production system are easy to use and to illustrate, these types of system become extremely complex as we relax the constraints and add new features. We are dealing here with infinite areas in state space, very high dimensional systems whose analysis is difficult and applications as yet unclear. Whilst in theory they have tremendous potential to model intelligent processes, they are at the cutting edge of complex systems research, especially when combined with the other complexity formalisms mentioned herein.

The ability to develop systems that can learn, that can modify themselves, that can act on their environment and also generate novelty opens up a future of intelligent machines with all the philosophical and technical problems that this brings. Whether this research direction will bear fruit or not is less important than the opportunity it gives us to understand the complexity of our own planet and the emergent creative interplay between diverse agents and diverse situations that can arise in strongly cooperating models.

8.7 In the field of Animation

Hansrudi NOSER in his PhD thesis titled as "A Behavioral Animation System Based on L-systems and Synthetic Sensors for Actors" elaborated the application of L-systems in animation and generation of synthetic sensors for actors [55]. Some important figures have been given for better visualization. By seeing all these figures a researcher may be tempted to generate many applications on animation and graphics. The following figure has been imported from this thesis.

Figure 8.1: Synthetic actors reproduced with permission from Hansrudi NOSER from his PhD thesis titled "A Behavioral Animation System Based on L-systems and Synthetic Sensors for Actors"

Some excerpts from this thesis have been mentioned for better understanding.

The creation or at least the understanding of living, autonomous and intelligent creatures has always been a challenge for scientists. Today, we have the possibility to experiment with primitive autonomous creatures in virtual environments simulated on powerful computers. By programming the actors with some behavioral rules one can study possibly meaningful emergent behaviors of individuals or societies. Through special interfaces a user can even interactively participate in a virtual scene. Autonomous synthetic actors can only exist in an environment. The way they perceive their environment influences drastically their behavior. In a behavioral animation system the three components consisting of the environment, the sensors of the creatures and the creatures themselves play an important role.

The aim of this thesis is to implement such an animation system in order to study behaviors and to create autonomous actors useful in film productions and interactive games. For this purpose, a formal theory of a behavioral L-system was developed. It is based on a timed, parametric, conditional, stochastic, context dependent and environmentally sensitive production system. The production system associates to its symbols some basic geometric primitives as cubes, spheres, trunks, cylinders terminated at their ends by half spheres, line segments, pyramids and imported triangulated surfaces. They defined non-generic environment elements as the ground plane, the sky, walls or other static objects directly in the axiom of the production system. The generic parts as growing plants are defined by production rules having only their germ in the axiom. The actors are also represented by special symbols. Their geometric representation can vary from some simple primitives like some cubes and spheres, over a more complicated skeleton structure to a fully deformed triangulated body surface usable in a ray tracer. It depends on the purpose of the applications that range from tests over interactive real time applications to ray traced video production.

An autonomous agent can capture this world by synthetic vision, sense of touch and audition. The synthetic vision is the simulated vision for a synthetic actor where the computer renders from the actor's point of view the virtual environment in a window corresponding to the actor's vision image. This Z-buffered and colored image permits an actor to recalculate the 3D position of pixels and to extract semantic data from the image. We implemented also vision sensor functions returning useful information from the vision system. These functions can be called in the condition of production rules defining, for example, collision avoidance behaviors.

To simulate sense of touch we defined functions used also in conditions of production rules evaluating the global force field at a given position. The force field function returns the amount of the global force field at the sensor position, and by comparing this value with a threshold value; a kind of collision detection with force field modeled environments can be simulated. A special sound event handler manages all sound events produced by an animation. To model an acoustic sensor for actors we defined a special function returning information about on-going sound events. This function can be used in the conditions of production rules to model sound event dependent behaviors of actors with production rules.

In their animation system they have several possibilities to define behavior. They described the group animation of fishes and butterflies with force fields defined in the L-system. Additionally, they presented an automata approach for defining behaviors based on synthetic vision, audition, and force field sensing and visual memory integrated in the L-system animation system. With the extensions of the behavioral L-system we can also define behaviors directly with production rules. With the above described sensor functions in the conditions of the production rules and the concept of the query symbol of environmentally sensitive L-systems simple rules can lead to interesting emergent behaviors.

The animation system is a universal, real time structured multi-L-system interpreter. Its real time structure allows interactive games with autonomous actors as partners. Additionally, it is conceived for automatic ray traced image-by-image film sequence production with synchronized sound track generation. Through a shared memory interface articulated humanoid actors with body deformation, controlled by a concurrent process can be imported into a simulation.

Another shared memory interface allows the sensor based autonomous actors to participate in the "Virtual Life Network". We demonstrate this feature with a networked interactive tennis game simulation where autonomous actors play the role of the referee judging the game and the role of a virtual game partner. The proposed animation system is a versatile research tool. It combines in a systemic approach several different components important and necessary for sensor based behavioral animation.

Nowadays computers enable us to create virtual environments offering a vast field of applications in domains as interactive computer games, interactive education, simulations, synthetic film productions, architecture and engineering. The role of such virtual environments becomes more and more important for our society in the next years. With virtual environments we try to reconstruct reality or to create fantastic imaginary worlds limited only by our imagination. A real environment consists of inanimate matter and living creatures as plants animals and humans obeying physical laws and behaving according to their inherent nature. Living creatures have the possibility to reproduce themselves, to grow and to react to their perceived environment. Even plants sense their environment and react to certain stimuli as light, gravitation, temperature and humidity. Animals and humans, in addition, can move around, and they show instinctive behaviors. Finally, humans are also equipped with some intelligence. Of course, with the present technical possibilities they could not exactly reproduce reality in a virtual environment. All they could achieve is a simplified model of a real environment. They tried to imitate certain principles and rules of reality in such a way that an immersed user can recognize it.

The objective of this work is to create a versatile; real time structured behavioral animation system allowing experimenting with virtual environments and autonomous agents. In reality, all autonomous agents are equipped with sensors through which they get information from the environment. Everything they know from their environment they learn through their sensors. This information severely influences their behavior. Thus, the first step in modeling "natural" behavior of autonomous agents is the introduction of synthetic sensors through which they will get all their environmental information. We will present sensor models for vision, audition and touch, which seem to us to be the most important senses for humanoid actors. During a real time structured animation an autonomous agent can only know the past, and at each time step it has to decide what to do next, based on its sensorial current input, its internal state and its knowledge. Thus, autonomous actors can be partners in interactive applications. Such a complex animation system is composed of several interacting sub-systems as geometric modeling, acoustic modeling, physical modeling, synthetic sensors or actor behaviors, for example. For its realization they adopted a systemic approach which views the animation system as an organized entity composed of interdependent elements, which are understood in terms of their relationships inside the global entity. According to computer science, artificial intelligence, ecology and psychology are privileged domains for systemic approaches as all of them are composed of interacting elements forming a complex entity.

In a behavioral animation system they first had to model the environment, the actors and the behavior of the actors before starting an animation. The classic description of 3D environments by polygons can be a very tedious work. To exploit fully the advantages of a computer and to facilitate the designing work of complex worlds we decided to develop a rewriting system having an important data amplification factor. In a rewriting system or a production system at each time step the environment is described by a symbol string generated by the computer according to a set of production rules. With such a production system we can symbolically describe not only the topology, but also the growth and behavior of complex articulated or branched objects. A text file of one or several pages containing these rules can produce several billions of bytes of images of a video film sequence. It can be considered as a compression of the film with a factor typically bigger than one million. In general, a production system needs not to build up a 3D database of the complete environment. It can directly draw objects during the interpretation of a symbolic environment string. Thus, the synthetic vision sensor approach for environmental information passing to the agents is an advantageous principle when using a production system.

In Computer Graphics image and film production is a necessity. Therefore, an animation system has to support it. Especially, film production by synthetic images is a tedious work for users of modeling and animation systems. Different programs with different scaling, orientation and data formats that have to be put together before the final rendering create very often parts of the 3D environment. Camera settings, key frame animation and sound tracks have to be coordinated. Very often the final product can only be seen after days of calculation revealing perhaps some errors or aesthetic problems that provoke a repetition of the whole procedure. That's why we paid special attention to an integration of an automatic film sequence production facility, allowing after a fast preview of the animation an automatic ray traced image-by-image video film production with synchronized sound track. Another objective is to enable a user of the software familiar only with L-systems to model and to produce easily film sequences without being a specialist for a dozen of other related software packages.

An important issue of this work is to be open to already existing or to future humanoid software at LIG. As these software libraries are already very complex, they decided to provide shared memory interfaces for inter process communication. This provides several advantages. First, the development can clearly be separated in smaller, easier compilable and maintainable units. Second, on multi - processor machines, as the Onyx from SGI, for instance, an application will turn much faster as parallelism is an inherent feature of this approach. Third, several different languages, such as C, C++ or perhaps Lisp in future development, can be used in the same project, and extensions to networked applications are easily implemented.

In all these different domains of behavioral animation a lot of specialized work has already been done. A major contribution of this work is that it combines many systems in a systemic approach to a new entity, a behavioral animation system suitable for studying and modeling synthetic sensor based humanoids in a convenient virtual environment. We think that its following features mainly give the originality of the present work:

- Development of a systemic, real-time structured animation system for sensor based humanoids.
- Control and modeling of the virtual environment (acoustic, geometric and force field model) by L-systems
- L-system extension with interacting particles, synthetic sensors (vision, touch, and audition) and behavior control through production rules and automata.
- Synthetic vision with visual memory and associated behaviors.

Real time structured sound rendering for film production (and in future for VR participants)

It supports interactive and networked partners mixed with autonomous sensor based actors. For example, a tennis game facility with an autonomous referee, with autonomous players and with interactive users as players has been developed.

They started with simple L-system modeled actors. Then, they passed to humanoid actors imported through a shared memory interface. Finally, we present the immersion of the autonomous actors of the L-system interpreter into the Virtual Life Network. After the description of the behavior control, the currently implemented automata based behaviors are presented. In particular they presented the L-system definition file format, some modules of the L-system interpreter, the speech recognition system and the image-by-image film production support offered by the animation system. In particular, they presented force field animations, plant, fractal and crystal modeling, navigation results and tennis game simulations.

Many interesting figures are generated using the concept discussed. Some of these figures are given below whose coloured version can be seen from the coloured section towards the end of the book.

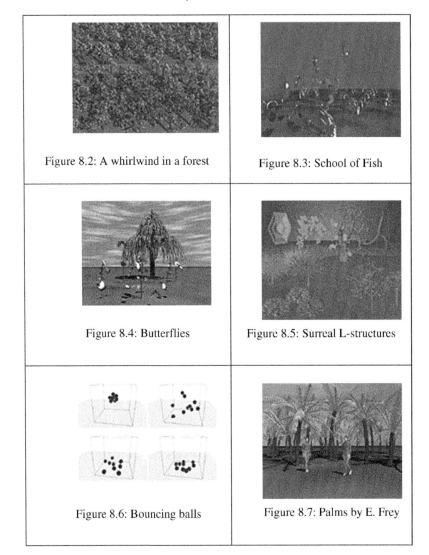

Figure 8.2: A whirlwind in a forest

Figure 8.3: School of Fish

Figure 8.4: Butterflies

Figure 8.5: Surreal L-structures

Figure 8.6: Bouncing balls

Figure 8.7: Palms by E. Frey

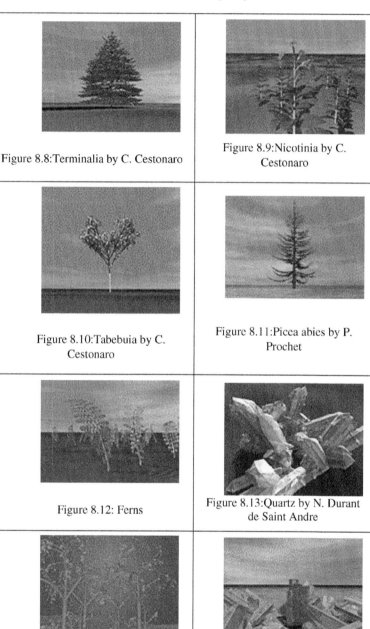

Figure 8.8:Terminalia by C. Cestonaro

Figure 8.9:Nicotinia by C. Cestonaro

Figure 8.10:Tabebuia by C. Cestonaro

Figure 8.11:Picea abies by P. Prochet

Figure 8.12: Ferns

Figure 8.13:Quartz by N. Durant de Saint Andre

Figure 8.14: Growing forest

Figure 8.15:Berillium by N. Durant de Saint Andre

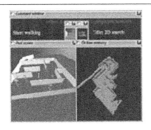

Figure 8.16:Vision based 2D navigation

Figure 8.17:Vision based 3D navigation

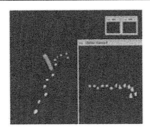

Figure 8.18: Walking on sparse foothold locations

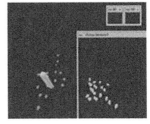

Figure 8.19: Walking on sparse foothold locations

Figure 8.20: Imported actor trajectories

Figure 8.21:Escaping from a maze

Figure 8.22:Walking through a flower field

Figure 8.23:Navigating humanoids

Figure 8.24: Actors of a tennis game

Figure 8.25:Tennis game

Figure 8.26:VLNET tennis stade

Figure 8.27:VLNET racket picking

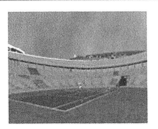

Figure 8.28:Autonomous referee and player

Figure 8.29:VLNET tennis game

Figure 8.30:Force field - tree interaction

Figure 8.31:Future animation with humanoids

Table 8.1: Figures reproduced with permission from Hansrudi NOSER from his PhD thesis titled "A Behavioral Animation System Based on L-systems and Synthetic Sensors for Actors"

8.8   Some important research abstracts

Evolving L-Systems To Generate Virtual Creatures *Computers and Graphics*, **25:6, pp 1041-1048. 2001.**
Hornby, G. S. and Pollack, Jordan. B.

Virtual creatures [43] play an increasingly important role in computer graphics as special effects and background characters. The artificial evolution of such creatures potentially offers some relief from the difficult and time-consuming task of specifying morphologies and behaviors. But, while artificial life techniques have been used to create a variety of virtual creatures, previous work has not scaled beyond creatures with 50 components and the most recent work has generated creatures that are unnatural looking. Here we describe a system that uses Lindenmayer systems (L-systems) as the encoding of an evolutionary algorithm (EA) for creating virtual creatures. Creatures evolved by this system have hundreds of parts, and the use of an L-system as the encoding results in creatures with a more natural look.

Fault-Aware Job Scheduling for Blue Gene/L Systems

A. J. Oliner, Massachusetts Institute of TechnologyR. K. Sahoo, IBM T.J. Watson Research CenterJ. E. Moreira, IBM T.J. Watson Research Center M. Gupta, IBM T.J. Watson Research CenterA. Sivasubramaniam, Pennsylvania State University

Large-scale systems like BlueGene/L are susceptible to a number of software and hardware failures that can affect system performance. In this paper evaluate the effectiveness of a previously developed job-scheduling algorithm for BlueGene/L in the presence of faults. We have developed two new job-scheduling algorithms considering failures while scheduling the jobs. We have also evaluated the impact of these algorithms on average bounded slowdown, average response time and system utilization, considering different levels of proactive failure prediction and prevention techniques reported in the literature. Our simulation studies show that the use of these new algorithms with even trivial fault prediction confidence or accuracy levels (as low as 10%) can significantly improve the performance of the BlueGene/L system.

The Application of L-systems and Developmental Models to Computer Art, Animation and Music Synthesis

Jon McCormack Degree: PhDYear Published: 2003

This thesis addresses the development of Lindenmayer systems (L-systems) and associated developmental models for the purposes of creating generative art, animation, and music. The research presented takes formalisms initially developed for the modelling and visualisation of biological systems and extends them as a creative tool for the generation of time-based visual and sonic structures. New techniques for the geometric interpretation of produced strings are introduced. These techniques improve over previous methods in their flexi-bility and scope in modeling organic form. Extensions introduced include: the use of generalised cylinders as a biologically inspired modelling tool, a variety of stochastic basis functions, and communication functions that connect the developing grammar with external events and environments. New developments are illustrated by applying them to modelling phyllotaxis, surfaces representing landscapes, and population distribution models over such surfaces. An analytical method for modelling phyllotaxis over arbitrary surfaces of revolution is presented. Temporal and developmental extensions specific to the generation of animated visual models and musical events are discussed in relation to L-systems. A number of related extensions are combined to form a generalised developmental model suited for hierarchical specification of both morphogenetic geometric models and musical events. Examples illustrate the application to the animation of articulated legged figures and the growth and development of plant models. Aesthetic evolution techniques for L-systems, and the application of object-oriented methods suited to synchronisation and real-time performance in practical implementations, are also described. These technical developments are placed in context historically by examining previous attempts at applying formal systems to artistic applications, and the use of botanical and biological visualisation in creative and scientific contexts. Specific topics on realism and mimesis in relation to computer graphics and views of nature and natural systems are considered. In particular, the concept of emergence is discussed in relation to generative art such as that produced by the formal systems detailed in this thesis.

Realistic Image Synthesis of Plant Structures for Genetic Analysis.
D. Wang, D. J. Kerbyson, G. J. King, and G. R. Nudd.. *Image and Vision Computing*, 19(8):517-522, May 2001.

The objective of this study was to develop a systematic method to restore images of plants using a small number of parameters in order to facilitate investigation of the genetic behaviour of branching patterns by computer models. Based on biological data, a generic methodology for image synthesis of plant structures was proposed. The corresponding algorithms for feature extraction and tuning were designed. Images of plant structures were synthesised with a higher accuracy by a model that was constructed using parametric L-systems. Comparing the synthetic images with the real plants validated the algorithms. The approach makes it possible to assess the virtual architectures of plants by computer models, instead of real experiments, and to investigate the relationship between plant morphology and their genetic behavious. This may aid the understanding of complex branching patterns that result from a plant's genetic behaviour.

Need and feasibility of applying L-system models in agricultural crop modeling
L. Pachepsky, M. Kaul, Ch. Walthall, J. Lyden, C. Daughtry (USA)

Contemporary agriculture uses crop simulation models for crop management and yield prediction. However, a model validation remains a permanent problem and interfaces for users contain only quantitative information. L-systems model coupled with a crop model could provide an additional visual validation of the latter and serve as an interactive and attractive for practical users interface. An open parametric L-systems model was developed for soybean based on the data for two cultivars growing in controlled climate chambers at three temperatures. Detailed morphological observations accompanied by frequently taken photographs were conducted from emergence to seed filling and were used to create the virtual plants. A functional model of vegetative development was parameterized for these two cultivars and linked with the visual model. The results showed that the virtual plants reproduced well the series of photographs of the real plants. Therefore, it can be useful for visual validating the phenological modules of crop models and as a part of user interface. The software L-Studio that was used for the visual modeling demonstrated extraordinary abilities. However, a serious obstacle for its extensive using in crop simulation is the absence of a manual understandable for a wide range of researchers working with crop models.

A structural method for assessing self-similarity in plants
Pascal Ferraro, Christophe Godin, and Przemyslaw Prusinkiewicz LaBRI, Université de Bordeaux INRIA, UMR Cirad-INRA-CNRS AMAP, Montpellier Department of Computer Science, University of Calgary

The important role of architecture in the understanding of plants generates a need for investigational tools. Generic tools have already been developed to visualize plant architecture in three dimensions, to model the development of plant structure, to measure plant architecture, and to analyse and quantify relations between plant components. This paper introduces a new tool for the identification of self-similarity within plant architectures.

Using L-Systems for Modeling the Architecture and Physiology of Growing Trees: The L-PEACH Model
Mitch Allen, Przemyslaw Prusinkiewicz, and Theodore DeJong Department of Pomology, University of California, Davis Department of Computer Science, University of Calgary

Carbohydrate partitioning represents a central problem of process-based models of tree growth because of the coupling between carbon partitioning, growth, and architecture. PEACH was an early, sink-driven, carbohydrate-partitioning model for simulating reproductive and vegetative growth of fruit trees. Carbon partitioning in that model was based on the hypothesis that a tree grows as a collection of semi-autonomous but interacting sinks (organs), and that these organs compete for resources. Organs of the same type were clustered into composite compartments, such as roots, fruit, or stems. Carbon was allocated to compartments depending on their competitive ability with respect to other compartments, and relative proximity to carbon sources. Biomass growth was dependent on an experimentally derived growth potential for each organ type. This approach made it possible to avoid the empirical allocation coefficients, functional balance rules, and allometric relationships that were common to most other tree models at the time. However, as pointed out by Le Roux et al., the PEACH model almost entirely ignored the interaction between tree architecture and carbon allocation. In addition, each organ type was treated collectively as a single compartment, and thus all organs of the same type grew at the average rate for that organ. Because of these limitations, there was no potential to simulate differences in organ size or quality as a function of location in the canopy. It was also impossible to use this model structure to simulate the function of individual organs and capture the influence of their performance on patterns of carbon partitioning. Overcoming these limitations requires a more detailed model of carbon economy, in which growth and function of each organ is modeled individually within an architecturally explicit model of canopy growth.

Simulation Modeling of Growing Tissues
Colin Smith and Przemyslaw Prusinkiewicz University of Calgary

Growing biological systems can be described as *dynamical systems with a dynamical structure*. In such systems not only the values of variables characterizing system components, but also the number of components and the connections between them, may change over time. For example, in a developing plant, the numbers of branches leaves and flowers change as the plant grows, and the topology and form of the plant are gradually modified. A simulation model of a plant must therefore be capable of dealing with the changing numbers of components, and their varying configurations. This problem was explicitly addressed by Lindenmayer, who introduced L-systems as formalism for modeling structures with a dynamically changing topology. This formalism, extended with a geometric interpretation, has subsequently been applied to model a wide variety of plants.

The L-system-based plant-modeling environment L-studio 4.0

Radoslaw Karwowski and Przemyslaw Prusinkiewicz Department of Computer Science, University of Calgary

L-studio and the Virtual Laboratory (Vlab) are two related plant modeling packages developed at the University of Calgary, Canada. This University has distributed them since 1999. The packages run on Windows and Linux machines, respectively, have similar functionality, and support the exchange of models between both systems. The main difference between these systems is the user interface, which is designed to meet the different expectations of Windows or Linux users. Below we outline the most recent version of L-studio (4.0).

Hairs, Textures, and Shades:Improving the Realism of Plant Models Generated with L-systems

Martin Fuhrer

High-quality, realistic visualization of plant models is a long-standing goal in computer graphics. Plants are often modeled using L-systems. Strings of symbols generated by the L-systems may be interpreted graphically as drawing commands to a rendering system. In this research, techniques for improving the appearance of plants generated from L-systems are proposed. A method of incorporating dynamic material specifications in L-system strings is presented, along with shading and lighting considerations for leaves and petals. Texture mapping of generalized cylinders is revisited in order to properly fit leaf and petal textures onto surfaces, and procedural methods for generating venation patterns and translucent rims on these surfaces are introduced. Finally, a method of generating hairs and controlling their parameters with L-systems is proposed. The importance of these techniques is illustrated in numerous state-of-the-art plant renderings.

Modeling Fracture Formation on Growing Surfaces

Pavol Federl

This thesis describes a framework for modeling fracture formation on differentially growing, bilayered surfaces, with applications to drying mud and tree bark. Two different, physically based approaches for modeling fractures are discussed. First, an approach based on networks of masses and springs is described. Two important shortcomings of the mass-spring approach are then identified: a tendency of the fractures to align themselves to the underlying mesh, and an unclear relationship between the simulation parameters and the real material properties. The rest of the thesis is focused on the second approach to modeling fractures, based on solid mechanics and finite element method, which effectively addresses the shortcomings of the mass-spring based approach. Two types of growth are investigated: isogonics and uniform anisotropic. The growth is then incorporated within the framework of finite element methods, implemented using velocity vector fields. The growth tensor is used to verify the properties of the growth. A number of efficiency and quality enhancing techniques are also described for the finite element based approach. An adaptive mesh refinement around fracture tips is introduced, which reduces the total number of elements required. Construction of a temporary local multi-resolution mesh around the fracture tips is described, which is used to calculate stresses at fracture tips with increased accuracy but without introducing any additional elements permanently into the model. Further, a general method for efficient recalculation of an equilibrium state is presented, which can be employed after localized changes are made to the model. Finally, an adaptive time step control is described, which automatically and efficiently determines the next optimal time step during simulation.

Improving the Process of Plant Modeling: The L+C Modeling Language
Radoslaw Karwowskiradekk@cpsc.ucalgary.ca

In this thesis he presented the modeling language L+C. L+C is a language based on the formalism of L-systems. It has been created to address the need for a formalism that would allow the expression of complex plant models. Current plant models require the components of the model (organs or cells) to include many parameters to describe the state of the model. Also the need to express complex calculations has been addressed.

Signal propagation has been traditionally expressed using context-sensitive L-systems. L+C extend the formalism of L-systems by introducing new concepts: derivation direction and new context. These two concepts are the foundation of a new method of propagating signals in plant models: fast information transfer. Fast information transfer is an alternative, faster method of propagating signals in linear and branching structures represented by L- system strings.

The L+C modeling language is implemented in a plant modeling program lpfg, which together with cpfg (another L-system-based modeling program developed at the University of Calgary) are the core part of the modeling environment L-studio.

Language-Restricted Iterated Function Systems, Koch Constructions, and L-systems

Przemyslaw Prusinkiewicz and Mark HammelDepartment of Computer Science University of Calgary Calgary, Alberta, Canada T2N 1N4

Linear fractals can be generated using a variety of methods. This raises the question of finding equivalent methods for generating the same fractal. Several aspects of this question have been addressed in the literature. These include:

A method for converting Koch constructions to equivalent iterated function systems (IFS's),

Methods for converting selected classes of L-systems with geometric interpretation to extensions of IFS's, such as controlled iterated function systems (CIFS's) and mutually recursive function systems (MRFS's).

Previous course notes introduced the notion of language-restricted iterated function systems (LRIFS's) encompassing CIFS's and MRFS's, and included a number of sample LRIFS's equivalent to L-systems with turtle interpretation. The present notes include the following further extensions to these results:

Introduction of the notions of iterated transformations of coordinate systems (ITCS's) and their language-restricted generalization (LRITCS's),

A characterization of Koch constructions in terms of ITCS's,

A method for constructing an LRITCS equivalent to a given LRIFS,

Expression of LRITCS's using parametric L-systems with turtle interpretation.

Readers are advised that this work is still in progress, and consequently the results are not presented with the rigor expected in final publication.

Growing Axons Evolving L-systems

A. Carrascal, D. Manrique, D. Pérez, J. Ríos, and C. Rossi (Spain)

They proposed a new technique for building neural networks which takes inspiration from the biology of neural cells. In our model neurons are fixed in space, and their connections are grown according to a given rule, for building any type of network without training. The growing rule is encoded in each cell's genetic code, and the strength of the connections depends on their length. Rules are generated using an evolutionary process. This process mimics the brains innate capabilities codifying this complex biological process with very few elements. Our first experimental results, both on laboratory tests and on a real world case, confirm the effectiveness of the proposed approach.

Representation of Fractals by means of L-Systems
Manuel Alfonseca

Fractals can be represented by means of L-systems (Development Grammars), together with a graphic interpretation. Two families of graphic interpretations have been used: turtle graphics and vector graphics. This paper describes an APL2/PC system able to draw fractals represented by L-systems, with both graphic interpretations. A theorem is proved on the equivalence conditions for both interpretations. Another point shown is the fact that supposed deficiencies in L-systems that have prompted proposals of extensions are really deficiencies in the graphic translation scheme.

Animation Based on the Interaction of L-Systems with Vector Force Fields

This paper discusses the use of rewriting systems for animation purpose. In particular, it describes the design of timed parameterized L-systems with conditional and pseudo-stochastic productions. It proposes a formulation to integrate the various features of L-systems into a unique L-system. It also introduces into the symbolism a way of using vector force fields to simulate interaction with the environment. The implementation, based on an object-oriented methodology, is described and animation examples are presented. Some images are extracted from this paper.

Figure 8.32 L-structures

Figure 8.33 L-System Tree

Figure 8.34 L-System Trees

Modeling the development of multicellular organisms using genetic L-systems
Przemyslaw Prusinkiewicz (University of Calgary)

While the modeling of genetic regulatory networks (GRN) in unicellular organisms is increasingly well understood, extensions to multicellular organisms continue to represent a methodological challenge. Genetic L-systems offer a method for modeling multicellular organisms with a branching topology. An extension of this notion is applicable to the modeling of cellular layers.

The formalism of L-systems was introduced in 1968 by A. Lindenmayer as a mathematical framework for describing the development of multicellular organisms. L-system models harness complexity of a developing organism by decomposing it into populations of modules. Depending on the model scale, these modules may represent cell components, cells, multicellular tissue regions, or higher-level architectural units, such as internodes, leaves, flowers, and buds in plants. As an organism develops, the number and configuration of its components change over time. L-systems automatize the process of dynamically updating the set of variables that characterize the organism state, and the system of equations that govern the state transitions. In genetic L-systems, the state variables include concentrations of proteins, and the equations include a mathematical model of a GRN. There are no restrictions on the formalism used to specify the GRN. For example, it may be described using Boolean networks, systems of differential equations, or Petri nets. He presented the notion of genetic L-systems using models of heterocyst differentiation in the blue-green bacterium Anabaena, and outline current work applying L-systems to the modeling of phyllotaxis and branching structure development in Arabidopsis according to molecular data. The development of these models represents an ongoing collaboration including biologists, computer scientists, and mathematicians. He illustrated using the interactive simulations implemented using the L-system-based simulation software L-studio, developed at the University of Calgary.

The Language LinF for Fractal Specification
Fernando M. Q. Pereira, Leonardo T. Rolla, Cristiano G. Rezende, Rodrigo L. Carceroni

This article presents the language LinF for L-System specifications. L-Systems are formal string rewriting systems introduced in 1968 by the botanist Aristid Lindenmayer that are used to model fractal images. LinF allows the definition of three-dimensional fractals and stochastic fractals. Together with the LinF definition this paper presents the implementation of cFleo, an OpenGL-based system that generates fractal images from LinF specifications. Results obtained through the LinF formalism show the ease of use and generality of the developed tools with respect to the existing literature.

Raytracing 3D Linear Graftals
Ph. Bekaert, Y. D. Willems
Dept. of Computing Science Katholieke Universiteit Leuven Celestijnenlaan 200A,3001 Leuven, Belgium

Many objects in nature, like trees, mountains and seashells, have a property called selfsimilarity. Sometimes this property is very pronounced, other natural phenomena exhibit this property to a lesser degree. During the past years much attention has been paid to fractals, purely self-similar objects. They present formalism, based on the well-known object-instancing graph, to represent objects, which are not necessarily purely selfsimilar. They show that Iterated Function Systems and some famous variants can be described elegantly in this formalism.

Organic and Semi-Organic Modelling Using Bracketed L-Systems
A CMPT461 final project proposal by *Jeremy Holman*

Bracketed L-Systems are noted for their ability to generate pleasing detailed models of organic systems, particularly plants, with desirable features including gradiated variability and an inherent ease of adaptation to chronological evolution. Many man-made systems, for example city planning, are a combination of designed and grown elements. I intend to implement L-Systems for PBRT, test this on plant structures as already seen in literature, and then experiment with using L-Systems for modelling semi-organic systems.

In any creative endeavour, sophisticated tools that do some of the work for the human artist or renderer are useful to have, even if they are inappropriate in some circumstances. In this vein, when we wish to use geometric models of objects pursuant to rendering them in synthetic image creation, in some (frequent but not ubiquitous) circumstances handling every detail manually is highly undesireable, and thus algorithmic generation of such models, or parts of models, is of interest. Backgrounds in animations are a good example of this, as our criteria for them are usually vastly under-constraining.

Along these lines, the observed self-similarity of many natural phenomena has lead to investigations of fractal and psuedo-fractal mathematical systems for this purpose, including Bracketed L-Systems, a type of formal grammar. The idea is that by modelling growth (eg plant growth) over time, via these relatively simple formal recursive relationships, we can generate surprisingly satisfying models, complete with diachronic histories should we want them.

In the first, Smith introduces the field, and explores the relationship of L-System-based model generation with some related fractal and psuedo-fractal approaches. Despite the simplicity of Smith's presentation, some of his images are surprisingly realistic and beautiful. Smith distinguishes between the genetic stage, which uses L-Systems, and which he considers to be graph-theoretic, and a phenotypic stage, when the L-Systems stop propgating and some interpretive rules are applied. At this early stage of my understanding, it seems to me that this distinction is the distinction between non-terminals and terminals (although L-Systems are not supposed to have terminals), and it further seems to me that his choice to put geometry (for example branching angles) exclusively in the second category is just that: a choice.

In the second, a single selection of a series of related papers by the authors, Prusinkewicz et al develop a refinement of the L-System approach which is inspired by the body of technique developed by 2d visual artists for representing plants. The advantages of this approach generally center around its greater intuitive access; it is easier to see how to manipulate the input parameters to meet certain desired gross constraints.

The work of Prusinkewicz et al is interesting, and he intended to pursue an implementation of it, the overall philosophy is synchronic rather than diachronic; descriptive rather than simulationist. He thinks that, generally speaking, the descriptive approach gives better results in known cases, but generalizes more poorly, and consequently he is also interested in the diachronic approach. Consequently, as he attempted to extend L-Systems to the modelling of semi-organic systems, balancing the benefits and requirements of these two approaches will be an ongoing issue.

He observed the challenge here as breaking down into two general areas: formal language challenges, and challenges in the implementational environment. The first challenge is to specify a well-defined grammar, and perform the necessary manipulations on it. The second challenge is to fit this specification into the PBRT input format, and to condition the results of the formal manipulations for expression in the language of PBRT's Shapes.

The PBRT input format already allows the definition of shapes ahead of time (using BeginShape(strings s)), which can then be instantiated later as appropriate. He intended to use this facility as a significant part of my framework. As a simple example, a leaf and a branch-segment might be modelled explicitly, stored using the BeginShape() facility, and then many copies of both assembled into a tree or shrub using a Bracketed L-System defined elsewhere. My understanding is that there is not a facility in PBRT for a similar BeginTransformation(), but only a transformation stack; he will look into the ease of implementing such functionality, because I currently suspect that such would be very helpful.

His aim is to have a robust L-System implementation for PBRT, which allows a user to specify some arbitrary primitives (some of them textured shapes or sets of textured shapes, others transformations), and also specify an L-System grammar, and then instantiate the grammar with the primitives and some other parameters (at least a number of generations) to generate at render-time a corresponding model and thus resultant image.

Once this is completed, he will experiment with applying this both to generation of entirely organic models, principally plants, and also to generation of semi-organic models, as inspiration presents itself.

Grammar-Based Music Composition

Jon McCormack
Computer Science Department Monash University, Clayton Victoria 3168

L-Systems have traditionally been used as a popular method for the modelling of space-filling curves, biological systems and morphogenesis. In this paper, we adapt string-rewriting grammars based on L-Systems into a system for music composition. Representation of pitch, duration and timbre are encoded as grammar symbols, upon which a series of rewriting rules are applied. Parametric extensions to the grammar allow the specification of continuous data for the purposes of modulation and control. Such continuous data is also under control of the grammar. Using non-deterministic grammars with context sensitivity allows the simulation of Nth-order Markov models with a more economical representation than transition matrices and greater flexibility than previous composition models based on finite state automata or Petri nets. Using symbols in the grammar to represent relationships between notes (rather than absolute notes) in combination with a hierarchical grammar representation, permits the emergence of complex music compositions from relatively simple grammars.

Music is one of those areas of activity that, despite its ubiquitous presence in human culture, remains largely immune to a detailed understanding. Nobody knows why something as apparently simple as the succession of changing discrete tones has the power to move the listener emotionally. Of all the arts, music is considered "to be open to the purest expression of order and proportion, unencumbered as it is by material media" It has been often noted that music bears a close affinity to mathematics, and that notions of mathematical and musical aesthetics may have some similarities. In recent years, many researchers and musicians have turned to the computer as both a compositional and synthesis device for musical expression.

In this paper, they describe a system for computer-assisted music composition and therefore the majority of this introduction will focus on the application of computers for composition. This is not to discount the many other uses of computers in music, such as sound synthesis, acoustic and physical modelling, automated notation, score editing and typography.

A major part of music creation involves the use of certain musical formalisms (systematic ordering), as well as algorithmic and methodic practices. Many of these were well established before the advent of digital computers. In a substantial survey of computer composition, Loy suggests that the application of musical formalisms, based on a priori theories of composition, have been largely developed this century, originating with composers such as Schoenberg and Hindemith. Loy also notes the contribution of music typographers and instructors who have found formal systems attractive to their profession.

Representations for notation and programming of music (and thus computer composition) can be broadly classified into symbolic and iconic. Symbolic representations relate symbols and a sonic event, but have no direct resemblance between them. Iconic representations carry some resemblance (usually visual) to the sound the icon represents. Common music notation combines both forms of representation: time signature, cleffs, bar lines are symbolic; pitch coding, crescendo, diminuendo are iconic. While this system is not completely adequate, particularly for some contemporary forms of music, it is an extremely versatile and capable notation for a diverse range of musical expression.

It has been observed that almost all forms of music involve repetition, either of individual sequences of notes or at some higher levels of structural grouping. Often these repetitions appear at a number of different levels simultaneously. Some compositions repeat patters that are slightly changed at each repetition, or revolve around some musical "theme" whereby complex harmonic, timing or key shifts in a basic theme provide a pleasing musical diversity.

It is also commonly known that what allows us to identify an individual piece of music is the change in pitch between notes, not the pitch of the notes themselves. We can change the key of a composition (equivalent to multiplying all frequencies by some fixed amount) and still recognise the melody.

A wide variety of approaches have been taken to the compositional problem. In Loy's survey paper, he details many different methods: stochastic and combinatorial models, grammar-based models, algorithmic models, process models and other models derived from Artificial Intelligence techniques.

The methods used in the system described in this paper are grammar-based. While grammars have been a popular method for algorithmic composition, in most instances, the grammars used are relatively primitive and thus limit the scope of control and diversity of composition that can be generated. The approach used in the system described in this paper, is to take recent developments in grammars used for modelling of biological morphogenesis and herbaceous plants in computer graphics, and apply them to music composition. In addition to being able to represent a variety of both stochastic and deterministic compositional techniques, the system has compositional properties unique to this approach.

The remainder of this paper is organised as follows. The next section briefly looks at the problem of computer based musical creativity and explains the rationale behind computer-assisted composition. The section "Representations for Music Composition" examines algorithmic representations for music composition and introduces the basis of the grammar method. The section "Grammars for Music Composition" details how grammars can be adapted for music composition, with specific emphasis on Lindenmayer Systems (L-Systems) as the basis for the model. The following section discusses implementation details. The final section details possible extensions and further work.

A complete theory of general creativity, or even musical creativity, remains elusive. While many attempts have been made to study and document the creative process - for example - any generality seems difficult to uncover. In many cases, people do not know how or why they make creative decisions, and much of the creative process is difficult to repeat in controlled experiments.

The question of whether creativity is computable is an issue of even greater controversy. Many researchers from Poincaré to Penrose have argued against a computable model of creativity because, simply put, the underlying mental processes are not computable. These arguments depend largely on speculation and the personal opinion of the authors (although Penrose does base his objections on an, as yet, untested theory of the relationship between consciousness and Quantum theory). It is this author's opinion that, while creativity is clearly an ability of the human mind, to a large extent practical creativity is deeply related to an individual's life experience. A life experience includes relationships to the environment, interaction with both living and nonliving things, social and cultural constructions. We all know that these things have a major effect on a person's internal states. Much creativity also depends on serendipitous and chance events in the external world, both conscious and unconscious. The explicit use of random events in composition is a practice many centuries old. It can be found in the works of many artists - from Mozart and Hayden, to Burroughs and Pollack. Creativity is not only influenced by external events, it may also be stimulated by artificially induced internal effects - some artists claim their best creative work is done under the influence of a large variety of chemical substances.

Some day it *may* be possible to build a computer with processing capabilities similar to that of the human brain. However, computing a simulated life experience would appear to be impossible. It is feasible that a robot that has life-experience in the real world may have the potential to exhibit creative behaviour but again, if life-experience is bound to physicality and matter (as opposed to mechanisms and processes) any creativity exhibited by such a robot may not be recognised as human-like creativity, or even recognised as creativity at all.

Johnson-Laird has argued that music is an excellent testing ground for theories of creativity because, as opposed to other Artificial Intelligence domains such as natural language generation, music avoids the problem of semantics; that is, a musical expression cannot be judged to be true or false. While this statement is literally correct, many musicians would argue that as part of the compositional process, musical expressions *do* contain semantics. For example, some composers associate emotive, structural or linguistic semantics with individual themes or passages in a composition. Carl Orff described musical composition as the raw expression of human energies and states of being. The musical development of a composition is driven by the associated development of emotions, structure or language. In this sense, such compositions do have "meaning" and there is a semantic association between musical expressions. Such expressions have ordering and denote an emotive symbolic system. According to Langer, music as a symbolic system of emotional archetypes conveys a "morphology of feeling" akin to algebraic notation conveying a mathematical expression.

Some composers (Brian Eno, for example) visualise abstract or real environments as mental images, and the visual and emotive feeling of these environments directly influences the composition. Composition is rarely a one-way process from mind to finished composition either. It is an *iterative feedback* process - hearing the music causes change in the composition. Composers are rarely interested in one aspect of the composition in isolation, such as pitch, timing or timbre; rather these things in total - the "surface" of the music - greatly influence the emotional feeling and thus carry the composition.

In summary, while music may appear to be an easier domain to test theories of creativity, it is difficult to expect a computer program to exhibit a genuine musical creativity, which approaches that of human composers. A more achievable approach, one adopted by the system described in this paper, is to use the machine as a synergetic partner to the human composer. The human and computer work in tandem through an interactive feedback process, the computer presenting new musical possibilities by synthesising complexity and variation and the human composer directing the overall creative "quality" of the composition. This human-machine feedback process, known as *computational synergetics* has been used to significantly enhance both scientific and artistic creativity. The computer, used in heuristic mode can lead the human collaborator to new discoveries that may have been difficult or impossible without the collaboration. In this way, the machine acts as a kind of "creative amplifier" enhancing the creative potential of the composer.

A grammar-based approach to composition makes heavy use of *procedural models*. One advantage of procedural methods in general, is that they allow a terse representation of a complex model. This is often referred to as *database amplification*. Such techniques are not the exclusive domain of computing. The representation of a complex organism is effectively encoded in the much simpler (in relative terms) DNA. Likewise, with grammars there is the possibility for the *emergence* of complexity through the repeated production of simple (and possibly deterministic) rules. One has to be careful with such analogies however, for biological morphogenesis relies also on many factors, such as the laws of physics, special properties of carbon atoms, the *self-organisational* properties of many molecular and cellular structures and complex environmental interactions. Certainly, these features are not built into our music compositional model, though it is possible that some of these concepts, in an abstract sense, could be incorporated into the model.

A diverse variety of *stochastic* processes have been applied to music composition. Most involve two stages. Firstly, some existing musical expression or quantity is analysed. Following the analysis, re-synthesis is applied using the selected stochastic technique. Importantly, while we can undertake statistical analysis of many musical patterns, when we attempt to synthesise music based on that analysis, the results rarely seem to have the clarity or intent of human composition. For example, Voss analysed many different types of music and found that several were statistically similar to *1/f* noise. However, when one converts *1/f* noise to musical notes, the results have little in common musically with any of the compositions from which the statistical analysis was derived. Given the discussion in the first section regarding formalisms, it is clear that a more sophisticated model is required.

Generally in composition, one primary concern is the notion of temporal *events* and *event spaces*. We use the general term *event* to represent some basic building block of composition - in many instances, a note or set of notes, but in other compositional applications, events may refer to sound complexes, individual sound samples or other basic musical elements. An *event space* is an ordered set of events. The number of events contained in a given event space represents its *order*.

One popular approach to computer composition is the use of *Markov chains*, where the probability of a specific event i is context-dependent on the occurrence of a previous event or event space. Most attempts using this form of modelling first require some existing composition to be analysed. For the purposes of explanation, we will restrict this discussion to consider only the pitch of notes. However, there is no reason why the same techniques cannot be applied to other musical qualities such as duration, volume or timbre. We also consider only *discrete-valued* (as opposed to *continuous-valued*) data. Markov processes are suited to the analysis of conventional musical pitches. A Markov process is considered *stationary* if the transition probabilities remain static.

An *N*-order Markov model can be represented using an *N+1* dimensional *transition matrix*. For example, Figure 1 in this paper shows a sequence of note events and the corresponding transition matrix. This is a 1st-order model, where the probability of a given event depends only on the event immediately preceding it. Each element in the transition matrix, *Pij* represents the probability of event *j* occurring given that event *i* has just occurred. *Pij* is calculated by summing the occurrences of each note *j* that follows note *i* in the input melody. Counts in each row are normalised so the combined probabilities for each row sum to 1.

$$\sum P_{ij} = 1 \forall_i \in [1..N]$$

The grammar-based approach is not new to music composition. Early users of grammar-based techniques include Buxton *et al.*   and Roads. Holtzman's Generative Grammar Definition Language (GGDL) compiler is based on Chomsky grammars of type 0 to 3 (regular, context-sensitive and context-free). Jones discusses context-free *space grammars* that can operate across many dimensions, where each dimension represents a different musical attribute. An interesting property of space grammars is that an *n*-dimensional grammar can generate an *n*-dimensional geometric shape, which may provide a higher level iconic representation of the music generated by a particular set of rules.

Little use of L-Systems has been made in the area of music composition. Prusinkiewicz has used two-dimensional space-filling curves generated by L-Systems as a basis for music generation, where the spatial position of a vertex in the curve represents a note, and the edge length, the duration.

Much development in music composition and synthesis has been performed in closely related areas such as fractal and stochastic methods, procedure-oriented and recursive composition. Moore gives an overview of some of these methods.

In techniques that use Chomsky grammars, the rewriting is *sequential*; that is, symbols are replaced one at a time sequentially across the string (corresponding to the serial nature of the process that the grammar represents). In the system described in this paper, we adapt *parallel* rewriting grammars for composition. Due to constraints discussed in the next section, rewriting is not fully parallel as is the case with L-Systems (upon which the system is based). Rewriting proceeds on parallel sets of elements in the string after the events represented by the string have occurred.

Development of the system is still in early stages, but already it is possible to see potential from the grammar-based approach. The use of stochastic grammars allows the compact representation Markov chains. The parallel rewriting technique of L-grammars has capabilities equal to, or better than, those of previous grammar-based methods. Parametric numerical parameters allow the control of non-discrete processes such as attack velocity or volume. Finally, hierarchical and context-sensitive grammars allow the simultaneous development of complex patterns at different levels within a composition.

In the introduction, it was emphasised that this is a computer-assisted composition system, where composer and machine work in a synergetic tandem. Currently, this tandem is little better than the way a programmer edits and debugs source code. Grammars must be written, compiled (removing syntax errors in the process) before music can be heard. Currently, work is being done to make the system much more interactive from a performance point of view. A composer will author a grammar and, within the grammar, place areas of discrete and continuous control that can be influenced by external processes or systems. One such example would be the recognition of hand gestures (gestures have both continuous and discrete information) in the spirit of the GROOVE system of Mathews and Moore. In this way the composer can act as conductor, instantiating medium-term changes in the performance through gesture (discrete gestures cause rule changes) and also influencing the short term quality of the performance (continuous data from hand positions in space). Development of such a system is currently underway.

It is also possible to apply "genetic" processes to grammars; to interactively evolve better sounding compositions using a novel variation of the genetic algorithm. Evolution of grammars has already been done for computer graphics modelling. Starting with a base grammar, a number of different mutated versions of the grammar can be represented as pictorial icons on the computer screen. By moving the mouse towards an icon, the contribution to the composition is increasingly biased towards the associated grammar. If the mouse is completely on the icon, only that grammar is heard. Once the user finds the best spacial position (and thus the best sounding composition), the selected grammars are "mated" and become the new parent grammar. The process is repeated for as long as the user wants, hopefully evolving the grammar into a better sounding composition.

While this technique sounds appealing, there is a limit to the number of mutations that can be displayed on the screen at any one time and also to the number of generations that can be critically evaluated by a human during any one sitting. We are still a long way from allowing the machine to exhibit creative judgment on its own.

Simulation of plant growth and eco-physiology by L-System based : Fractal generated : Turtle interpreted computer graphics model

Jiang Xiangning, Wang Tianhua, Chen Xuemei, and Gag Xiaoyi The Experimental Center of Forest Biology, College of Plant SciencesBeijing Forestry University, 100083, Beijing, China.

An L-System-based: Fractal-generated: Turtle-interpreted: OOP techniques-implemented computer graphics model (LFT) has been encoded for theoretical plant growth and eco-physiology study. The model is divided into four function modules for parameter input, plant growth and forest stand modelling, results output, and virtual experiment design and conduct. Based on input parameters abstracted from field experiments and theoretical constants, the model preliminarily can graphically and dynamically simulate plant/tree growth and their stands, calculate leaf area index (LAI) of a stand, and determine optimal leaf density in a defined space and light interception ratio, etc. for plant growth, physiology, ecology, and theoretical biology study. Modelling results can be put out as a data table, line/bar chart, and/or graphics.

Arterial Branching within the Confines of Fractal L-System Formalism

M. Zamir Department of Applied Mathematics and Department of Medical Biophysics. University of Western Ontario, London, N6A 5B7, Canada

Parametric Lindenmayer systems (L-systems) are formulated to generate branching tree structures that can incorporate the physiological laws of arterial branching. By construction, the generated trees are de facto fractal structures, and with appropriate choice of parameters, they can be made to exhibit some of the branching patterns of arterial trees, particularly those with a preponderant value of the asymmetry ratio. The question of whether arterial trees in general have these fractal characteristics is examined by comparison of pattern with vasculature from the cardiovascular system. The results suggest that parametric L-systems can be used to produce fractal tree structures but not with the variability in branching parameters observed in arterial trees. These parameters include the asymmetry ratio, the area ratio, branch diameters, and branching angles. The key issue is that the source of variability in these parameters is not known and, hence, it cannot be accurately reproduced in a model. L-systems with a random choice of parameters can be made to mimic some of the observed variability, but the legitimacy of that choice is not clear.

Visualization of developmental processes by extrusion in space-time
Mark Hammel and Przemyslaw Prusinkiewicz

Developmental processes in nature may involve complex changes in the topology, shape, and patterns of growing structures. Processes taking place in one or two dimensions can be visualized as objects in three-dimensional space, obtained by extruding the growing structures along a line or curve representing the progress of time. In this paper, we extend the notion of L-systems with turtle interpretation to facilitate the construction of such objects. This extension is based on the interpretation of the entire derivation graph generated by an L-system, as opposed to the interpretation of individual words. We illustrate the proposed method by applying it to visualize the development of compound leaves, a sea shell with a pigmentation pattern, and filamentous bacteria. In addition to serving as visualization examples, these models are of interest on their own. The sea shell model uses an L-system to express a reaction-diffusion process, thus relating these two models of morphogenesis. The model of bacteria, which is also of the reaction-diffusion type, sheds new light on one of the basic problems of morphogenesis, the formation of equally spaced organs in a developing medium.

Representation of Fractal Curves by Means of L Systems
Manuel Alfonseca and Alfonso

Fractals can be represented by means of L-systems (Development Grammars), together with a graphic interpretation. Two families of graphic interpretations have been used: turtle graphics and vector graphics. This paper describes an APL2/PC system able to draw fractals represented by L-systems, with both graphic interpretations. A theorem has been proved on the equivalence conditions for both interpretations. Another point shown is the fact that supposed deficiencies in L-systems that have prompted proposals of extensions are really deficiencies in the graphic translation scheme.

Recognition of plants using a stochastic L-system model, pp.50-58
Ashok Samal, Univ. of Nebraska/Lincoln, Lincoln, NE,USA;Brian Peterson, Univ. of Nebraska/Lincoln, Lincoln,NE, USA;David J. Holliday, Univ. of Nebraska/Lincoln, Lincoln, NE, USA.Journal of Electronic Imaging Journal of Electronic Imaging, January 2002, Vol. 11(01)

Recognition of natural shapes like leaves, plants, and trees, has proven to be a challenging problem in computer vision. The members of a class of natural objects are not identical to each other. They are similar, have similar features, but are not exactly the same. Most existing techniques have not succeeded in effectively recognizing these objects. One of the main reasons is that the models used to represent them are inadequate themselves. In this research we use a fractal model, which has been very effective in modeling natural shapes, to represent and then guide the recognition of a class of natural objects, namely plants. Using the stochastic L-systems accommodates variation in plants. A learning system is then used to generate a decision tree that can be used for classification. Results show that the approach is successful for a large class of synthetic plants and provides the basis for further research into recognition of natural plants.

Fractal Modeling of Structure and Dynamical Pattern Formation in Brain Neural Systems using Nonlinear Equations with Derivatives of Fractional Order

Vladimir Khasilev Barrington University, Boca Raton, FL

Modeling of the growth and functioning of neuron sets revealed relationship between fractal space structure and fractal time function. The goal of this paper is to provide a new mathematical approach to describe this relationship using nonlinear partial differential equations with derivatives of fractional order. A recursive deterministic algorithm has modeled the space structure of the neural systems. The resulting neuron network may be subdivided into three subsets. L-subset (Logic) consists of comparatively short dendrites. The growth of these dendrites is deterministic with predictable result. The I-subset (Intuition) consists of comparatively long dendrites. This growth of this subset is deterministic as well, however unpredictable due to deterministic chaos. The X-subset consists of dendrites that do not reach any axon. This subset receives no input from own neural systems and, like all systems with sensory deprivation, is extremely sensitive to external signals such as electromagnetic and acoustic waves. Analysis of patterns formation reveals that the L-system possesses dynamical patterns (sequences of states) with transition between close states. The L-system with inserted I-system is capable of making sudden leaps between patterns with limited number of possible strategies. The L-system with inserted both I- and X-systems possess practically unlimited number of the strategies for the leaps between patterns. Possible application of the obtained results for the systems of artificial intellect with intuition is considered.

8.9 Research Proposal on L-System

Proposal-I: The issue of *parsing* has been considerably neglected in the study of parallel grammars. This is of course important if these concepts should be applied to model linguistic phenomena, which seems to be possible as well as reasonable in many ways. For example, observe that our brain is obviously working in a sort of parallel fashion, so that we might also assume that human syntax analysis of natural languages is actually performed in a parallel way. Closely related and important to this issue is the question of *ambiguity* in parallel grammars in a broad sense. Note that an ambiguous grammar offers various ways of "reading" or "interpreting" a given "sentence," which is largely unwanted if certain semantics is connected to each such interpretation.

Proposal-II: Another issue in this respect is a systematic study of *semantics*. To develop such a "parsing program," it is also important to give good automata.

Proposal-III: By interpreting L-System developments graphically in the turtle sense as described in Chapter 4, using appropriate rescaling, we may come to certain fractal limit objects. Such objects can also be described by mutually recursive function systems. Fractal parameters like the Hausdorff dimension of the described fractal can be computed under certain conditions.

Proposal-IV: It is quite imaginable that partial parallelism can be successfully applied in other classification tasks. For example, detecting similarities in trees in parallel could be used for supporting Internet browsing or finding illegal copies of software code, even after making obvious modifications like renaming variables. Partial parallelism or (probably equivalently) regulated rewriting can be used to keep track of non-local information.

Proposal-V: A better way for representing parallel grammar may be the use for regular expression, which may be useful for designing a complete compiler based on parallel grammar.

8.10 Summary

Fractal images generated using L-System concept, is relatively new and has been proved challenging. The field of research is far from being exhausted since there are many directions that have not yet been fully investigated (e.g., the use of non-affine transformations, image compression using L-System, use of 3D L-Systems, fractal encoding and the use of hybrid fractals discussed in this book etc.). We hope the directions highlighted in this chapter will definitely help many researchers to add the basic outlook in their interest in the field of L-System. All most every fractal can be regenerated using L-System concept. Particularly the growth phenomenon can be easily simulated by this concept with less mathematical complexity with least computational resources.

BIBLIOGRAPHY

1. A. Fournier, D. Fussel and L. Carpenter, Computer rendering of stochastic models. Communications of the ACM, 25(6), June, 1982, 371-384.
2. A. Habel and H.-J. Kreowski. On context-free graph languages generated by edge replacement. In H. Ehrig, M. Nagl, and G. Rozenberg, editors, *Graph grammars and their application to computer science; Second International Workshop*, Lecture Notes in Computer Science 153, pages 143–158. Springer-Verlag, Berlin, 1983.
3. A. Habel and H.-J. Kreowski. May we introduce to you: Hyperedge replacement. In H. Ehrig, M. Nagl, G. Rozenberg, and A. Posenfeld, editors, *Graph grammars and their application tocomputer science; Third International Workshop*, Lecture Notes in Computer Science 291, pages 15–26. Springer-Verlag, Berlin, 1987.
4. A. K. Bisoi and J. Mishra, On calculation of fractal dimension of images, *Pattern Recognition Letters*, 22, 2001, 631-637.
5. A. K. Bisoi and J. Mishra, Fractal images with inverse replicas, *Machine Graphics & Vision*, 8(1), 1999, 77-82.
6. A. K. Bisoi and J. Mishra, *Generation of fractal pattern with the help of a modified MRCM*, Proceedings of Conference Image and Vision Computing, New Zealand, IVCNZ98, 1998, 50-54.
7. A. K. Bisoi, S. N. Mishra and S. Mishra, *Generation of L-System string from ramification matrix*, Proceedings of Conference on Information Technology, CIT2005, Dec 20-23, Bhubaneswar, Orissa, 2005, 119-124.
8. A. K. Bisoi, S. N. Mishra and J. Mishra, Growing a class of fractals based on the combination of classical fractal and recursive mathematical series in L-system, *Machine Graphics & Vision*, 13(3), 2004, 275-288.
9. A. K. Bisoi, S. N. Mishra, S. Meher and S. Mishra, *A study of fractal figures by L systems using turtle graphics for designing a simple L-system Parser*, Proceedings of Conference on Information Technology: Prospects and Challenges in the 21st century *Kathmandu*, Nepal, May 23 - 26, 2003.
10. A. K. Bisoi, S. N. Mishra and S. Mishra, *Generation of self similar graphical images using iterated function system*, Proceedings of Conference on Information Technology, December 21-24, Bhubaneswar, Orissa, 2002, 300-301.
11. A. Lindenmayer. An introduction to parallel map generating systems. In H. Ehrig, M. Nagl, A. Rosenfeld, and G. Rozenberg, editors, *Graph grammars and their application to computer science; Third International Workshop*, Lecture Notes in Computer Science 291, pages 27–40. Springer-Verlag, Berlin, 1987.
12. A. Lindenmayer, Developmental systems without cellular interactions, their languages and grammars, *Journal of Theoretical Biology,* 30, 1971,455-484.
13. A. Lindenmayer, Mathematical models for cellular interactions in development II, Simple and branching filaments with two-sided inputs. *Journal of Theoretical Biology,* 18, 1968, 300-315.
14. A. Lindenmayer, Mathematical models for cellular interactions in development I, Filaments with one-sided inputs, *Journal of Theoretical Biology,* 18, 1968, 280-299.
15. A. Lindenmayer and G. Rozenberg, editors. *Automata, languages, development.* North-Holland, Amsterdam, 1976.

16. A. L. Szilard and R. E. Quinton. An interpretation for DOL systems by computer graphics. *The Science Terrapin*, 4, 1979, 8-13.

17. A. Norton, Generation and rendering of geometric fractals in 3-D, Computer Graphics, 16(3), 1982, 61-67.

18. A. Norton, Julia sets in the quaternions, Computers and Graphics, 13(2), 1989, 267-278.

19. A. R. Smith, Plants, fractals, and formal languages. Computer Graphics, 18(3), July, 1984, 1-10.

20. B. B. Chaudhuri and N. Sarkar, Texture segmentation using fractal dimension, *IEEE Trans. on Pattern Anal Machine Intell,* 17(1), 1995, 72-77.

21. B. B. Mandelbrot, Self-affine fractal sets, in *Fractal in Physics* (L. Pietronero & E. Tosatti Eds.), W.H.Freeman, Amsterdam, 1986.

22. B. B. Mandelbrot, *The Fractal Geometry of Nature*, W.H.Freeman, San Fransisco, 1982.

23. B. B. Mandelbrot, Comment on computer rendering of stochastic models. Communications of the ACM, 25(8), 1982, 581-583.

24. B. B. Mandelbrot, *Fractals: Form, Chance, and Dimension.* W.H. Freeman, San Francisco, 1977.

25. B. B. Mandelbrot and J. W. V. Ness, Fractional Brownian Motions, Fractional Noise and Applications, SIAM Review, 10(4), 1968, 422-437.

26. B. Dubuc, J. −F Quiniou, C. Roques-Crames, C. Tricot and S. W. Zucker, Evaluating the fractal dimension of profiles, *Phys. Rev. A.*, 39, 1989, 1500-1512.

27. B. J. West and A. L. Goldberger, Physiology in Fractal Dimensions, *American Scientist*, 75, 1987, 345-365.

28. C. A. Pickover and A. L. Khorasani, Fractal characterization of speech waveforrm graphs, *Computer Graphics* 1, 1, 1986, 51-61.

29. C. Bouville, Bounding ellipsoids for ray-fractal intersection, *Computer Graphics*, 19(3), 1985, 45-55.

30. C. R. Cook and P. S. P. Wang, A Chomsky hierarchy of isotonic array grammars and languages, *Computer Graphics and Image Processing,* 8, 1978, 144-152.

31. D. Ermel, k-limitierte IL-systeme, Studienarbeit, Technische Universitat Braunschweig, 2001, Studienarbeiten/Ermel.pdf.

32. D. J. Sandin, J. C. Hart L. H. Kauffman, Interactive visualization of complex, stacked and quaternion Julia sets, In Proceedings of Ausgraph '90, 1990.

33. D. Watjen, Parallel communicating limited and uniformly limited OL systems. *Theoretical Computer Science,* 255(1-2), 2001, 163-191.

34. E. Csuhaj-Varjii, J. Dassow, J. Kelemen and Gh. Paun, *Grammar Systems: A Grammatical Approach to Distribution and Cooperation.* London, Gordon and Breach, 1994.

35. F. K. Musgrave, C. E. Kolb and R. S. Mace, The synthesis and rendering of eroded fractal terrains. Computer Graphics, 23(3), July, 1989, 41-50.

36. G. Dong, Grammar tools and characterizations.   In *Proceedings of the Eleventh ACM SIGACT-SIGMOD-SIGART Symposium on Principles of Database Systems (PODS'92),* ACM, June, 1992, 81-90.

37. G. Lescinski, Lively IFS. SIGGRAPH Video Review, 60, (Animation), 1991.

38. Gh. Paun and A. Salomaa, editors, *Grammatical Models of Multi Agent Systems,* chapter H. Fernau, R. Freund, and M. Holzer: Regulated array grammars of finite index, pages 157-181 (Part I) and 284-296 (Part II). London: Gordon and Breach, 1999.

39. G. Rozenberg and A. Salomaa, editors, *L Systems,* volume 15 of *LNCS,* chapter A. Walker: Adult languages of L systems and the Chomsky hierarchy, pages 201-215. Berlin: Springer, 1994.

40. G. Rozenberg, TOL systems and languages. *Information and Control (now Information and Computation),* 23:357-381, 1973.

41. G. Rozenberg and A. Salomaa, editors, *Handbook of Formal Languages (3 volumes).* Springer, 1997.

42. G. Rozenberg and A. Salomaa. *The mathematical theory of Lsystems.* Academic Press, New York, 1980.

43. G. S. Hornby and B. Jordan, Evolving L-Systems To Generate Virtual Creatures, *Computers and Graphics,* 25(6), 1041-1048, 2001.

44. G. S. P. Miller, The definition and rendering of terrain maps. Computer Graphics, 20(4), August, 1986, 39-48.

45. G. Tankard, *Practical Modelling of Realistic Plants for a Real-Time 3D Environment,* Thesis for Bachelor of Science (Honours) degree at Rhodes University, 2001.

46. G. T. Herman and G. Rozenberg. *Developmental systems and languages.* North-Holland, Amsterdam, 1975.

47. Gy. Vaszil, On parallel communicating Lindenmayer systems, in Gh. Paun and A. Salomaa, editors, *Grammatical Models of Multi-Agent Systems,* volume 8 of *Topics in Computer Mathematics,*Gordon and Breach, Amsterdam, 1999, 99-112.

48. Gy. Vaszil, Communication in parallel communicating Lindenmayer systems, *Grammars,* 1(3), Special Issue on Grammar Systems, 1999, 255-270.

49. G. Y. Gardner, Simulation of natural scenes using textured quadric surfaces. Computer Graphics, 18(3), July 1984, 11-20.

50. H. Abelson and A. A. diSessa, *Turtle Geometry,* MIT Press, 1982.

51. H. Bordihn and M. Holzer, On the number of active symbols in L and CD grammar systems, *Journal of Automata, Languages and Combinatorics,* 6(4), 2001, 411-426.

52. H. Fernau, Closure properties of ordered languages, *EATCS Bulletin,* 58, February, 1995, 159-162.

53. H. Honda and J. B. Fisher. Tree branch angle: Maximizing effective leaf area. *Science,* 199, 1978, 888–890.

54. H. Honda and J. B. Fisher. Ratio of tree branch lengths: The equitable distribution of leaf clusters on branches. *Proceedings of the National Academy of Sciences USA,* 76(8):3875–3879, 1979.

55. H. Noser, A Behavioral Animation System Based on L-systems and Synthetic Sensors for Actors, PhD Thesis.

56. H. Sudborough, On tape-bounded complexity classes and multihead finite automata. *Journal of Computer and System Sciences,* 10(1), 1975, 62-76.

57. H. O. Pitmen, H. Jurgens and D. Saupe, *Fractals for Classrooms,* Springer, 1992.

58. J. A. R. Holbrook, Quaternionic astroids and starfields. Applied Mathematical Notes, 8(2), 1983, 1-34.

59. J.C.Hart, *The Object Instancing Paradigm for Linear Fractal Modeling*, Proceedings of Graphics interface '92 Vancouver, BC, 11-15 May 1992, 224-231.

60. J. C. Hart, *Image space algorithms for visualizing quaternion Julia sets*, Master's thesis, EECS Dept., University of Illinois at Chicago, 1989.

61. J. C. Hart and T. A. DeFanti, Efficient antialiased rendering of 3-D linear fractals, *Computer Graphics*, 25(3), 1991.

62. J. C. Hart, L. H. Kauffman and D. J. Sandin, *Interactive visualization of quaternion Julia sets*, In A. Kauffman, editor, Proceedings of Visualization '90, IEEE Computer Society, 1990, 209-218.

63. J. D. Foley, A. van Dam, S. K. Feiner and J. F. Hughes, *Computer Graphics: Principles and Practice. Systems Programming Series.* Addison-Wesley, 2nd edition, 1990.

64. J. Feder, *Fractals,* Plenum Press, New York, 1989.

65. J. Feng, W. -C. Lin and C. -T. Chen, Fractional box-counting approach to fractal dimension estimation, in *Proceedings ICP'R 96,* 1996, 854-858.

66. J. Gonczarowski and M. K. Warmuth, Scattered versus context-sensitive rewriting. *Acta Informatica,* 27, 1989, 81-95.

67. J. Gangepain and C. Roques-Carmes, Fractal approach to two dimensional and three dimensional surface roughness, *Wear,* 109, 1986, 119-126.

68. J. Hutchinson, Fractals and self-similarity. Indiana University Mathematics Journal, 30(5), 1981, 713-747.

69. J. M. Keller, S. Chen, R. M. Crownover, Texture description through fractal geometry, *Computer Vision Graphics Image Processing.*, 45, 1989, 150-166.

70. J. M. Keller, R. M. Crownover and R. Y. Chen, Characteristics of natural scenes related to the fractal dimension, *IEEE Trans. Part. Anal Machine Intell* 9, 5, 1987, 621-627.

71. J. Theiler, Estimating fractal dimension, *Journal of Optical Society of America.,* A7, 1990, 1055-1073.

72. J. T. Kajiya, New techniques for ray tracing procedurally defined objects, ACM Transactions on Graphics, 2(3), 1983, 161-183, Also appeared in Computer Graphics 17, 3, 1983, 98-102.

73. J. W. Carlyle, S. A. Greibach and A. Paz, A two-dimensional generating system modeling growth by binary cell division (preliminary report). In *XV. Annual conterence on swit. automata theory,* 1974, 1-12.

74. K. Culik II and S. Dube, L-Systems and Mutually Recursive Function Systems, *Acta Informat*, 30, 1993, 279-302.

75. K. Culik II and A. Lindenmayer, Parallel graph generating and graph recurrence systems for multicellular development. *International Journal of General Systems,* 3, 1976, 53-66.

76. K. J. Falconer, The dimensions of self-affine fractals II, *Math. Proc.* Camb. *Phil. Soc.,* 111, 1992, 169-179.

77. K. J. Falconer and B. Lammering, Fractal properties of generalised Sierpinski Triangles, *Fractals.*, 6(1), 1998, 31-41.

78. K. J. Falconer, *Fractal Geometry: Mathematical Foundations and Applications*, John Wiley & Sons, 1997.

79. K. J. Falconer and D. T. Marsh, The dimensions of affine-invariant fractals, *J. Phys. A.,* 21, 1988, L121-L125.

80. K. J. Falconer, *The Geometry of Fractal Sets.* Cambridge University Press, New York, 1985.

81. K. Perlin, An image synthesizer. Computer Graphics, 19(3), July 1985, 287-296.

82. K. Rozenberg and A. K. Salomaa, Developmental systems with fragmentation. *International Journal of Computer Mathematics,* 5, 1976, 177-191.

83. L. Linsen, B. J. Karis, E. G. McPherson and B. Hamann, *Tree Growth Visualization,* Proceedings of the Conference WSCG 2005, Plzen, Czech Republic, January 31-February 4, 2005.

84. L. Rosenberg, On multi-head finite automata. *IBM Journal of Research and Development,* 10, 1966, 388-394.

85. M. Alfonseca, and A. Ortega, A Study of Representation of Fractal Curves by L Systems and their equivalences, *IBM J.Res. & Dev.,* 41, 1997.

86. M. Alfonseca, and A. Ortega, Determination of Fractal Dimensions from Equivalent L-Systems, *IBM J. Res. & Dev.,* 45(6), 2001, 797-805.

87. M. de Does and A. Lindenmayer, *Algorithms for the generation and drawing of maps representing cell clones,* volume 153 of *LNCS,* Berlin, Springer, 1983, 39-57.

88. M. F. Barnsley, *Fractals Everywhere.* Academic Press, New York, 1988.

89. M. F. Barnsley, A. Jacquin, F. Mallassenet, L. Rueter and A. D. Sloan, Harnessing chaos for image synthesis. *Computer Graphics,* 22(4), 1988, 131-140.

90. M. Kudlek, Dead word languages and adult languages. *Pure Mathematics and Applications. Ser. A,* (3/4), 1990, 343 - 355.

91. M. K. Biswas, T. Ghose, S. Guha, P. K. Biswas, Fractal dimension estimation for texture images: A parallel approach, *Pattern Recognition Letters.* **19**, 1998, 309-313.

92. M. K. Levitina, On some grammars with rules of global replacement (in Russian). *Scientific-technical information (Nauchno-tehnichescaya informacii). Series 2,* (3), 1972, 32—36.

93. N. Nirmal and K. Krithivasan, Filamentous systems with apical growth. *International Journal of Computer Mathematics,* 12, 1983, 203-215.

94. N. Sarkar and B. B. Chaudhuri, An efficient differential box-counting approach to compute fractal dimension of image, *IEEE Trans. on Systems, Man. and Cybernetics,* 24(1), 1994, 115-120.

95. O. Deussen, P. Hanrahan, B. Lintermann, R. Mech, M. Pharr and P. Prusinkiewicz, Realistic modeling and rendering of plant ecosysterns. In *SIGGRAPH '98. Proceedings of the 25th Annual Conference on Computer Graphics,* ACM, 1998, 275-286.

96. O. Ibarra, Simple matrix languages. *Information and Control (now Information and Computation),* 17, 1970, 359-394.

97. P. Asvestas, G. K. Matsopoulos and K. S. Nikita, Estimation of fractal dimension of images using a fixed mass approach, *Pattern Recognition Letters.,* 20, 1999, 347-354.

98. P. Eichhorst and W. J. Savitch. Growth functions of stochastic Lindenmayer systems. *Information and Control,* 45:217–228, 1980.

99. P. Grassberger, Generalizations of the Hausdorff dimension of fractal measures, *Phys. lett.*, 107A, 1985, 101-105.

100. P. Linz, An Introduction to Formal Languages and Automata, Narosa Publishing House, New Delhi, India, 2nd Edition, 1997.

101. P. Pentland, Shading into texture, *Artificial Intell.*, 29, 1986, 147-170.

102. P. Pentland, Fractal based description of natural scenes, *IEEE Trans. Pattern Anal. Machine Intell.* 6, 1984, 661-674.

103. P. Prusinkiewicz, Graphical Applications of L-Systems, Proceedings of Graphics Interface 86 and Vision Interface 86, M. Wein and E. M. Kidd, Eds, Vancouver, BC, 1986, 247-253.

104. P. Prusinkiewicz and J. Hanan, *Lindenmayer Systems, Fractals, and Plants,* volume 79 of Lecture Notes in Biomathematics. Springer-Verlag, New York, 1989.

105. P. Prusinkiewicz and L. Kari, Subapical bracketed l-systems, in J. Curry, H. Ehrig, G. Engels, and G. Rozenberg, editors, *Grammars and Their Application to Computer Science,* 1073 of *LNCS,* 1996, 550-564.

106. P. Prusinkiewicz and A. Lindenmayer, *The Algorithmic Beauty of Plants,* Springer, New York, 1990.

107. P. Prusinkiewicz, M. Hammel and R. Měch, Visual Models of Morphogenesis : A Guided Tour, http://www.cpsc.ucalgary.ca/projects/bmv/vmm/

108. P. Prusinkiewicz, M. Hammel, R. Mech and J. Hanan, The artificial life of plants. In *Artificial Life for Graphics. Animation, and Virtual Reality,* volume 7 of *SIGGRAPH'95 Course Notes,* pages 1-38. ACM, ACM Press, 1995.

109. P. Prusinkiewicz, M. Hammel and R. Měch, *The Artificial Life of Plants,* Department of Computer Science, University of Calgary, 1995.

110. P. Prusinkiewicz, A. Lindenmayer and J. Hanan, Developmental models of herbaceous plants for computer imagery purposes, *Computer Graphics*, 22(4), 1988, 141-150.

111. P.S.P. Wang, Parallel context-free array grammar normal forms, *Computer Graphics and Image Processing,* 15, 1981, 296-300.

112. P.M.B. Vitanyi, Development, growth and time. In G. Rozenberg and A. K. Salomaa, editors, *The Book of L,* Berlin, Springer, 1985, 431-444.

113. R. Cretzburg, A. Mathias, E. Ivanov, Fast algorithm for computing the fractal dimension of binary images, *Physica-A*. **185**, 1992, 56-60.

114. R. Freund, Aspects of n-dimensional Lindenmayer systems, in G. Rozenberg and A. Salomaa, editors, *Developments in Language Theory; At The Crossroads of Mathematics. Computer Science and Biology (Turku. Finland. 12-15 July 1993),* Singapore: World Scientific, 1994, 250-261.

115. R. F. Voss, *Fractals in nature: From characterization to simulation,* In Peitgen, H. and Saupe, D., editors, The Science of Fractal Images, Springer-Verlag, New York, 1988, 2t-70.

116. R. F. Voss, Random fractals: Characterization and the Measurement, in *Scaling phenomena in Disordered Systems* (by R. Pynn and A. Skjeltorp, Ed.), Plenum, New York, 1985.

117. R. Siromoncy, On equal matrix languages, *Information and Control (now Information and Computation),* 14, 1969, 133-151.

118. R. Vollmar, *Algorithmen in Zellularautomaten. volume* 48 of *LAMM,* Stuttgart, Teubner, 1979.

119. S. Buckzkowski, S. Kyriacos, F. Nekka and L. Cartilier, The modified box-counting method: Analysis of some characteristic parameters, *Pattern Recognition.*, 31, 4, 1998, 411.418.

120. S. Chen. J. M. Keller and R. Crownover, On the calculation of fractal features from images, *IEEE Trans. Patt. Anal. Machine Intell.*, 15(10), 1993, 1087-1090.

121. S. Chen, J. M. Keller and R. Crownover, Shape from fractal geometry, *Artificial Intell.*,43, 1990, 199-218.

122. S.D. Casey and N.F. Reingold, Self-Similar Fractal Sets: Theory and Procedure, *IEEE Computer Graphics & Applications*, 14, 1994, 73-83.

123. S. Demko, L. Hodges and B. Naylor, Construction of Fractal Objects with Iterated Function Systems, *Computer Graphics,* 19(3), July 1985, 271-278.

124. S. Greibach and J. Hopcroft, Scattered context grammars. *Journal of Computer and System Sciences,* 3, 1969, 233-247.

125. S.Mishra, A Computer Graphics Study of Characteristics of Fractals, PhD Thesis, Utkal University, Vani Vihar, Bhubaneswar, India, 2006

126. S. Papert, Mindstorms: Children, Computers, and Powerful Ideas, *Basic Books,* New York, 1980.

127. T. Balanescu, M. Gheorghe and Gh. Paun, Three variants of apical growth; filamentous systems, *International Journal of Computer Mathematics,* 22, 1987, 227-238.

128. T. Yokomori. Stochastic characterizations of EOL languages. *Information and Control*, 45:26–33, 1980.

129. U. G. Gujar, V. C. Bhavsar, S. Y. M. Choi and P. K. Kalra, Traversed Geometric Fractals, *IEEE Computer Graphics & Applications*, September 1993, 61-67.

130. V. Aladyev, On the equivalence of t_m-grammars and Sb (n) grammars (in Russian). Commentationes Mathematicae Universitatis Carolinae, 15, 1974, 717-726.

131. V. Diekert and G. Rozenberg, editors, *Book of Traces,* World Scientific, Singapore, 1995.

132. V. Rajlich, *Absolutely parallel grammars and two-way deterministic finite state transducers*, in Third Annual ACM Symp.Theory Computing. ACM, 1971, 132-137.

133. Y. Termonia and Z. Alexandrowc, Fractal dimension of strange attractors from radius versus size of arbitrary clusters, *Phys Rev. Lett*, 51, 1983, 1265-1268.

134. Yi-Ping, Phoebe Chen and R. M. Colomb, Database technologies for L-system simulations in virtual plant applications on bioinformatics, *Knowledge and Information Systems*, 5(3), Springer-Verlag New York, Inc.

135. Yi-Ping and Phoebe Chen, *DML: A Bridge between Database Systems and L-Systems for Biological Research*, Proceedings of the 11th International Conference on Scientific on Scientific and Statistical Database Management, IEEE Computer Society, July 1999.

136. X. C. jin, S. H. Ong, A practical method for estimating fractal dimension, *Pattern Recognition Letter*, 16, 1995, 457-464.

APPENDIX-A

Few examples of L-System codes for Developed fractals

```
Koch1 { ; S N Mishra
  Angle 6
  Axiom F--F--F
  F=F+F--F+F
  }


Koch2 { ; S N Mishra
  Angle 12
  Axiom F---F---F---F
  F=-F+++F---F+
  }


Koch3 { ; S N Mishra
  Angle 4
  Axiom F-F-F-F
  F=F-F+F+FF-F-F+F
  }


Koch6 { ; S N Mishra
  axiom f+f+f+f
  f=f-ff+ff+f+f-f-ff+f+f-f-ff-ff+f
  angle 4
   }


Dragon { ; S N Mishra
  Angle 8
  Axiom FX
  F=
  y=+FX--FY+
  x=-FX++FY-
  }


Peano1 { ; S N Mishra
Angle 4
  Axiom F-F-F-F
  F=F-F+F+F+F-F-F-F+F
  }
```

```
Cesaro { ; S N Mishra
 Angle 34
 Axiom FX
 F=
 X=----F!X!+++++++++F!X!----
 }

DoubleCesaro { ; S N Mishra
 Angle 4
 axiom D\90D\90D\90D\90
 D=\42!D!/84!D!\42
 }

FlowSnake { ; S N Mishra

 angle=6;
 axiom FL
 L=FL-FR--FR+FL++FLFL+FR-",
 R=+FL-FRFR--FR-FL++FL+FR",
 F=
 }

CantorDust { ; S N Mishra
 Angle 6
 Axiom F
 F=FGF
 G=GGG
 }

Snowflake2 { ; S N Mishra
 angle 12
 axiom F
 F=++!F!F--F--F@IQ3|+F!F--
 F=F--F!+++@Q3F@QI3|+F!F@Q3|+F!F
 }

SnowflakeColor { ; S N Mishra
 angle 12
 axiom F
 F=--!F<1!F<1++F<1++F<1@IQ3|-F<1!F<1++
 F=F<1++F<1!---@Q3F<1@QI3|-F<1!F<1@Q3|-F<1!F<1
 F=
 }
```

```
Island1 { ; S N Mishra
  angle 4
  axiom F+F+F+F
  F=FFFF-F+F+F-F[-GFF+F+FF+F]FF
  G=@8G@18
  }


Island2 { ; S N Mishra

  angle 4
  axiom f+f+f+f
  f=f+gf-ff-f-ff+g+ff-gf+ff+f+ff-g-fff
  g=@6G@16
  }

Quartet { ; S N Mishra
  angle 4
  axiom fb
  A=FBFA+HFA+FB-FA
  B=FB+FA-FB-JFBFA
  F=
  H=-
  J=+
  }

SnowFlake1 { ; S N Mishra
  Angle 12
  Axiom FR
  R=++!FRFU++FU++FU!---@Q3FU|-@IQ3!FRFU!
  U=!FRFU!|+@Q3FR@IQ3+++!FR--FR--FRFU!--
  F=
  }

SnowFlake3 { ; S N Mishra
  angle 12
  axiom fx
  x=++f!x!fy--fx--fy|+@iq3fyf!x!+++f!y!+++f!y!fx@q3+++f!y!fx
  y=fyf!x!+++@iq3fyf!x!+++f!x!+++f!y!fx@q3|+fx--fy--fxf!y!++
  f=
  }
```

Tree1 { ; S N Mishra
 angle=12;
 axiom +++FX
 X=@.6[-FX]+FX
 }

Peano2 { ; S N Mishra
 Angle 8
 Axiom FXY++F++FXY++F
 X=XY@Q2-F@IQ2-FXY++F++FXY
 Y=-@Q2F-@IQ2FXY
 }

Sierpinski1 { ; S N Mishra
 angle 3
 axiom F
 F=FXF
 X=+FXF-FXF-FXF+
 }

Koch4 { ; S N Mishra
 angle 12
 axiom f++++f++++f
 f=+f--f++f-
 }

Plant07 { ; S.N.Mishra
 axiom Z
 z=zFX[+Z][-Z]
 x=x[-FFF][+FFF]FX
 angle 14
 }

Plant08 {
 axiom SLFFF
 s=[+++Z][---Z]TS
 z=+H[-Z]L
 h=-Z[+H]L
 t=TL
 l=[-FFF][+FFF]F
 angle 20
 }

```
Hilbert { ;
 axiom x
 x=-YF+XFX+FY-
 y=+XF-YFY-FX+
 angle 4
 }

Sierpinski3 {
 axiom F-F-F
 f=F[-F]F
 angle 3
 }


Peano3 {
 axiom x
 x=XFYFX+F+YFXFY-F-XFYFX
 y=YFXFY-F-XFYFX+F+YFXFY
 angle 4
 }

Koch5 {
 axiom f+F+F+F
 f=F+F-F-FFF+F+F-F
 angle 4
 }

Sierpinski2 { ;
 axiom FXF--FF--FF
 f=FF
 x=--FXF++FXF++FXF--
 angle 6
 }

SierpinskiSquare {
 axiom F+F+F+F
 f=FF+F+F+F+FF
 angle 4
 }
```

Pentagram { ; angle 10
  axiom fx++fx++fx++fx++fx
; f=f[++++@1.618033989f]
  x=[++++@i1.618033989f@.618033989f!x!@i.618033989f]
  }

QuadKoch { ; S N Mishra,
  angle 4
  AXIOM F-F-F-F-
  F=F+FF-FF-F-F+F+FF-F-F+F+FF+FF-F
  }

Fass1 { ; S N Mishra,
  axiom -l
  angle 4
  L=LF+RFR+FL-F-LFLFL-FRFR+
  R=-LFLF+RFRFR+F+RF-LFL-FR
  }

Fass2 { ; S N Mishra,
  angle 4
  axiom -l
  L=LFLF+RFR+FLFL-FRF-LFL-FR+F+RF-LFL-FRFRFR+
  R=-LFLFLF+RFR+FL-F-LF+RFR+FLF+RFRF-LFL-FRFR
  }

QuadGosper { ; S N Mishra,
  angle 4
  axiom -Fr
  l=FlFl-Fr-Fr+Fl+Fl-Fr-FrFl+Fr+FlFlFr-Fl+Fr+FlFl+Fr-FlFr-Fr-
Fl+Fl+FrFr-
  r=+FlFl-Fr-Fr+Fl+FlFr+Fl-FrFr-Fl-Fr+FlFrFr-Fl-FrFl+Fl+Fr-Fr-
Fl+Fl+FrFr
  f=
  }

Plant01 { ; S N Mishra,
  angle 14
  axiom f
  f=F[+F]F[-F]F
  }

```
Plant02 { ; S N Mishra,
  angle 18
  axiom f
  f=F[+F]F[-F][F]
  }

Plant03 { ; S N Mishra,
  angle 16
  axiom f
  f=FF-[-F+F+F]+[+F-F-F]
  }

Plant04 { ; S N Mishra,
  angle 18
  axiom x
  X=F[+X]F[-X]+X
  F=FF
  }

Plant05 { ; S N Mishra,
  angle 14
  axiom x
  X=f[+X][-X]FX
  F=FF
  }

Plant06 { ; S N Mishra,
  angle 16
  axiom x
  X=F-[[X]+X]+F[+FX]-X
  F=FF
  }

Plant09 { ; S N Mishra
  axiom y
  x=X[-FFF][+FFF]FX
  y=YFX[+Y][-Y]
  angle 14
}
```

Plant10 { ; S N Mishra
  axiom f
  f=f[+ff][-ff]f[+ff][-ff]f
  angle 10
  }

Plant11 { ; S N Mishra
  axiom f
  f=F[+F[+F][-F]F][-F[+F][-F]F]F[+F][-F]F
  angle 12
  }

Curve1 { ; S N Mishra,
  angle 4
  axiom F-F-F-F-
  f=FF-F-F-F-F-F+F
  }

Curve2 { ; S N Mishra,
  angle 4
  axiom F-F-F-F-
  f=FF-F+F-F-FF
  }

Curve3 { ; S N Mishra,
  axiom F-F-F-F-
  angle 4
  F=F-FF--F-F
  }

Curve4 { ; S N Mishra
  axiom yf
  x=YF+XF+Y
  y=XF-YF-X
  angle 6
  }

Leaf1 { ; S N Mishra,
  angle 8
  axiom x
  a=n
  n=o

```
o=p
p=x
b=e
e=h
h=j
j=y
x=F[+A(4)]Fy
y=F[-B(4)]Fx
F=@1.18F@i1.18
}

Leaf2 { ; S N Mishra,
 angle 8
 axiom a
 a=f[+x]fb
 b=f[-y]fa
 x=a
 y=b
 f=@1.36f@i1.36
}

Bush { ; S N Mishra
 Angle 16
 Axiom ++++F
 F=FF-[-F+F+F]+[+F-F-F]
}

MyTree { ; S N Mishra
 Angle 16
 Axiom ++++F
 F=FF-[XY]+[XY]
 X=+FY
 Y=-FX
}

ColorTriangGasket { ; S N Mishra
 Angle 6
 Axiom --X
 X=++FXF++FXF++FXF>1
 F=FF
}
```

SquareGasket { ; S N Mishra
 Angle 4
 Axiom X
 X=+FXF+FXF+FXF+FXF
 F=FF
 }

DragonCurve { ; S N Mishra
 Angle 4
 Axiom X
 X=X-YF-
 Y=+FX+Y
 }

Square { ; S N Mishra
 Angle 4
 Axiom F+F+F+F
 F=FF+F+F+F+FF
 }

KochCurve { ; S N Mishra
 Angle 6
 Axiom F
 F=F+F--F+F
 }

Penrose1 { ;
 Angle 10
 Axiom +WF--XF---YF--ZF
 W=YF++ZF----XF[-YF----WF]++
 X=+YF--ZF[---WF--XF]+
 Y=-WF++XF[+++YF++ZF]-
 Z=--YF++++WF[+ZF++++XF]--XF
 F=
}

 Angle 10
 Axiom +WC02F--XC04F---YC04F--ZC02F
 W=YC04F++ZC02F----XC04F[-YC04F----WC02F]++
 X=+YC04F--ZC02F[---WC02F--XC04F]+
 Y=-WC02F++XC04F[+++YC04F++ZC02F]-
 Z=--YC04F++++WC02F[+ZC02F++++XC04F]--XC04F

```
F=
}

Penrose2 { ;
 Angle 10
 Axiom ++ZF----XF-YF----WF
 W=YF++ZF----XF[-YF----WF]++
 X=+YF--ZF[---WF--XF]+
 Y=-WF++XF[+++YF++ZF]-
 Z=--YF++++WF[+ZF++++XF]--XF
 F=
}

Penrose3 { ;
 Angle 10
 Axiom [X]++[X]++[X]++[X]++[X]
 W=YF++ZF----XF[-YF----WF]++
 X=+YF--ZF[---WF--XF]+
 Y=-WF++XF[+++YF++ZF]-
 Z=--YF++++WF[+ZF++++XF]--XF
 F=
}

Penrose4 { ;
 Angle 10
 Axiom [Y]++[Y]++[Y]++[Y]++[Y]
 W=YF++ZF----XF[-YF----WF]++
 X=+YF--ZF[---WF--XF]+
 Y=-WF++XF[+++YF++ZF]-
 Z=--YF++++WF[+ZF++++XF]--XF
 F=
}

DoublePenrose { ;
 Angle 10
 Axiom [X][Y]++[X][Y]++[X][Y]++[X][Y]++[X][Y]
 W=YF++ZF----XF[-YF----WF]++
 X=+YF--ZF[---WF--XF]+
 Y=-WF++XF[+++YF++ZF]-
 Z=--YF++++WF[+ZF++++XF]--XF
 F=
}
```

```
Sphinx { ;
 Angle 6
 Axiom X
 X=+FF-YFF+FF--FFF|X|F--YFFFYFFF|
 Y=-FF+XFF-FF++FFF|Y|F++XFFFXFFF|
 F=GG
 G=GG
 }

PentaPlexity {
 Angle 10
 Axiom F++F++F++F++F
 F=F++F++F|F-F++F
 }

; old PentaPlexity:
; Angle 10
; Axiom F++F++F++F++Fabxjeabxykabxyelbxyeahxyeabiye
; F=
; a=Fabxjea
; b=++F--bxykab
; x=++++F----xyelbx
; y=----F++++yeahxy
; e=--F++eabiye
; h=+++++F-----hijxlh
; i=---F+++ijkyhi
; j=-F+jkleij
; k=+F-klhajk
; l=+++F---lhibkl

CircularTile { ; S N Mishra
 axiom
X+X+X+X+X+X+X+X+X+X+X+X+X+X+X+X+X+X+X+X+X+X+X+
X
 x=[F+F+F+F[---X-Y]+++++F++++++++F-F-F-F]
 y=[F+F+F+F[---Y]+++++F++++++++F-F-F-F]
 angle 24
 }

Lars1{ ;
 Angle 8  ; angle increment/decrement is 45
 axiom [F]++[F]++[F]++F
 F=F[+F][-F]
```

```
}

Lars2{ ;
 Angle 8
 axiom +[F]++[F]++[F]++F
 F=F[+F][-F]
 }

Lars1Color{ ;

 Angle 8  ; angle increment/decrement is 45
 axiom C1[F]++[F]++[F]++F
 F=F<1[+F][-F]>1
 }

Lars2Color{
 Angle 8  ; angle increment/decrement is 45
 axiom C1+[F]++[F]++[F]++F
 F=F<1[+F][-F]>1
 }

Man {
 ; looks like man with an odd number of iterations
 Angle 8
 Axiom F++F++F++F
 F=-F-FF+++F+FF-F
}

Lace { ;
 Angle 8
 Axiom F++F++F++F
 F=F+++F---F+F---F++F--F++F
}
```

APPENDIX-B

Few examples of L-System codes for various hybrid fractals

1. L-System for Table 4.11 of Chapter-4

```
Koch{
  Dirs = 25
  Axiom = A
  #Iterations = 3
  A=AF--F--F|BF+F--F+F
  };Koch

Koch1{
  Dirs = 6
  Axiom = F-F-F-F
  #Iterations = 3
  F=F-F+FF-F-F+F
};Koch1

Koch2{
  Dirs = 15
  Axiom = A
  #Iterations = 3
  A=AF-F-F|BF[-F]F
  };Koch2

Koch3{
  Dirs = 20
  Axiom = A
  #Iterations = 3
  A=AF-F-F-F|BF-F+F+FF-F-F+F
  };Koch3

Koch4{
  Dirs = 31
  Axiom = A
  #Iterations = 3
  A=AF+F+F+F|BF-FF+FF+F+F-F-FF+F+F-F-FF+F
};Koch4
```

2. L-System codes for Table 4.15 of Chapter-4

```
KochSuffixOneton{
  Dirs = 6
  Axiom = AF--F--F
  #Iterations = 3
  A=BF+F--F+F+F
  B=BF+F--F+F+FX
  X=
};KochSuffixoneton

KochSuffixOdd{
  Dirs = 6
  Axiom = AF--F--F
  #Iterations = 3
  A=BF+F--F+F+F
  B=BF+F--F+F+FXX
  X=
};KochSuffixOdd

KochPrefixOneton{
  Dirs = 6
  Axiom = F--F--FA
  #Iterations = 3
  A=F+F--F+F+FB
  B=F+F--F+F+FBX
  X=
};KochPrefixoneton

KochSubstiOneton{
  Dirs = 6
  Axiom = AF--F--F
  #Iterations = 3
  A=B
  B=BF+F--F+F+FX
  X=
};KochSubstiOneton

Koch1SuffixOneton{
  Dirs = 6
  Axiom = AF-F-F-F
  #Iterations = 3
```

```
  A=BF-F+FF-F-F+F
  B=BF-F+FF-F-F+FX
  X=
};Koch1SuffixOneton

Koch1PrefixOneton{
  Dirs = 6
  Axiom = F-F-F-FA
  #Iterations = 3
  A=F-F+FF-F-F+FB
  B=F-F+FF-F-F+FBX
  X=
};Koch1PrefixOneton

Koch1SubstiOneton{
  Dirs = 6
  Axiom = AF-F-F-F
  #Iterations = 3
  A=B
  B=BF-F+FF-F-F+FX
  X=
};Koch1SubstiOneton
```

APPENDIX-C

Few examples of L-System codes for chapter-5 table 5.4

```
OnetonKochSuffix{
  Dirs = 6
  Axiom = AF--F--F
  #Iterations = 1
    A=AF--F--FBF+F--F+F
  };OnetonKochSuffix

GrammarKoch{
  Dirs = 14
  Axiom = AF--F--F|BF+F--F+F|AF--F--F
  #Iterations = 1
  A=AF--F--F|BF+F--F+F|AF--F--F
  };GrammarKoch

Koch{
  Dirs = 6
  Axiom = F--F--F
  #Iterations = 1
  F=F+F--F+F
  };Koch

GrammarKoch{
  Dirs = 15
  Axiom = AF--F--F|BF+F--F+F|AF--F--F
  #Iterations = 1
  A=AF--F--F|BF+F--F+F|AF--F--F
  };GrammarKoch


GrammarKoch{
  Dirs = 12
  Axiom = AF--F--F|BF+F--F+F|AF--F--F
  #Iterations = 1
  A=AF--F--F|BF+F--F+F|AF--F--F
  };GrammarKoch
```

GrammarKoch{
  Dirs = 6
  Axiom = AF--F--F|BF+F--F+F|AF--F--F
  #Iterations = 1
  A=AF--F--F|BF+F--F+F|AF--F--F
  };GrammarKoch

GrammarKoch{
  Dirs = 24
  Axiom = AF--F--F|BF+F--F+F|AF--F--F
  #Iterations = 1
  A=AF--F--F|BF+F--F+F|AF--F--F
  };GrammarKoch

GrammarKoch{
  Dirs = 3
  Axiom = AF--F--F|BF+F--F+F|AF--F--F
  #Iterations = 1
  A=AF--F--F|BF+F--F+F|AF--F--F
  };GrammarKoch

GrammarKoch{
  Dirs = 21
  Axiom = AF--F--F|BF+F--F+F|AF--F--F
  #Iterations = 1
  A=AF--F--F|BF+F--F+F|AF--F--F
  };GrammarKoch

Colour Images

1. Table 3.4 : Figures generated by using blue prints

BLUE PRINT	GENERATED FIGURES
 The Sun (Blue print)	 The Sun
 Leaves (Blue print)	 Leaves
 The Goddess Subhadra (Blue print)	 The Goddess Subhadra

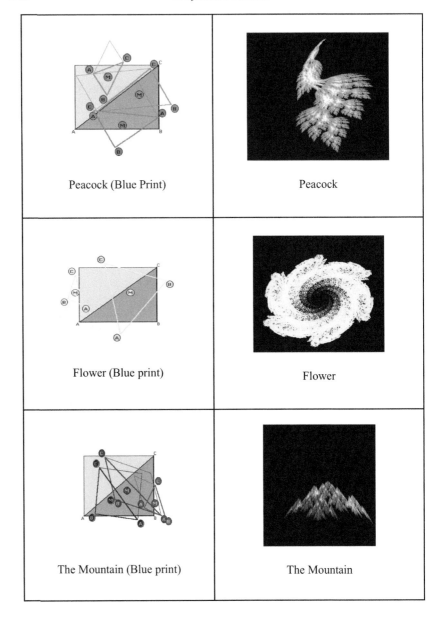

Peacock (Blue Print)

Peacock

Flower (Blue print)

Flower

The Mountain (Blue print)

The Mountain

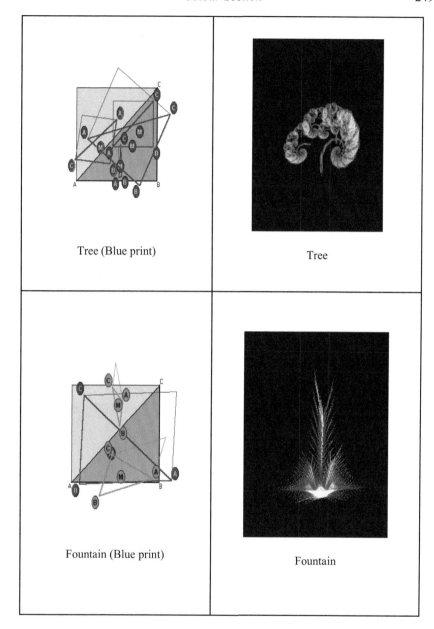

Tree (Blue print)

Tree

Fountain (Blue print)

Fountain

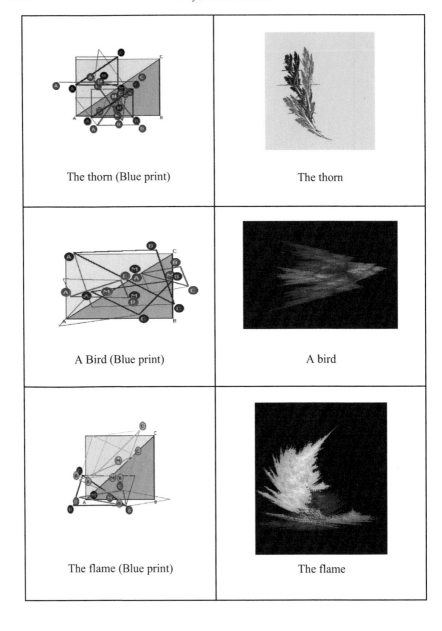

The thorn (Blue print)	The thorn
A Bird (Blue print)	A bird
The flame (Blue print)	The flame

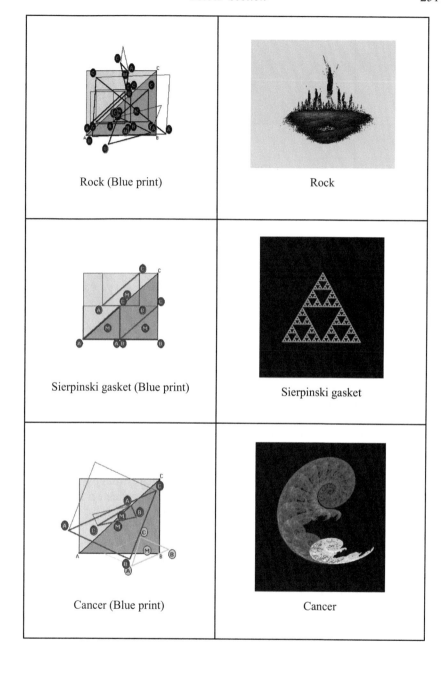

Rock (Blue print)

Rock

Sierpinski gasket (Blue print)

Sierpinski gasket

Cancer (Blue print)

Cancer

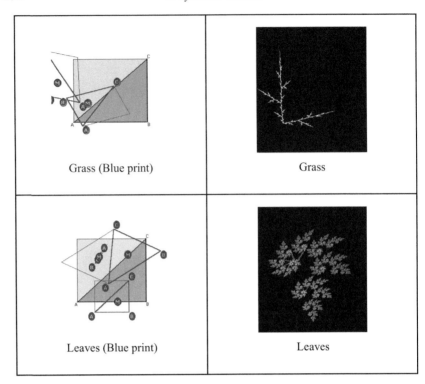

As discussed in chapter 3, beautiful and colourful fractals can b
generated out of different equations, represented through a blue print o
the concerned fractal. The table of colourful fractal images are shown i
the table above with corresponding blue prints against them. This ha
been reproduced with colours from the paper written by the authors title
"Generation of self similar graphical images using iterated functio
system" from the Proceedings of Conference on Information Technolog
Bhubaneswar, Orissa, 2002, 300-301 [10].

2. Table 8.1: Figures from the thesis of Hansrudi NOSER

Many interesting figures have been generated using the concept discusse
in the PhD thesis of H. NOSER titled "A Behavioral Animation Syster
Based on L-systems and Synthetic Sensors for Actors" in the area c
application of L-systems in animation and generation of synthetic sensor
for actors [55]. These colourful figures have been given in the tabl
below as a token of illusion and perception about 3D animation. This ha
been discussed in chapter 8 section 8.7 in detail.

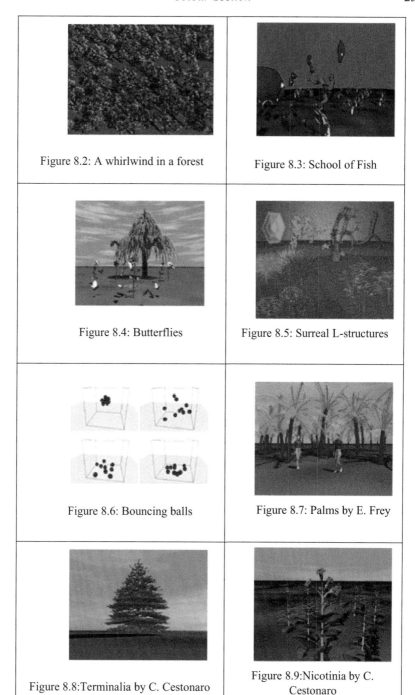

Figure 8.2: A whirlwind in a forest

Figure 8.3: School of Fish

Figure 8.4: Butterflies

Figure 8.5: Surreal L-structures

Figure 8.6: Bouncing balls

Figure 8.7: Palms by E. Frey

Figure 8.8:Terminalia by C. Cestonaro

Figure 8.9:Nicotinia by C. Cestonaro

Figure 8.10:Tabebuia by C. Cestonaro

Figure 8.11:Picea abies by P. Prochet

Figure 8.12: Ferns

Figure 8.13:Quartz by N. Durant de Saint Andre

Figure 8.14: Growing forest

Figure 8.15:Berillium by N. Durant de Saint Andre

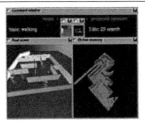

Figure 8.16:Vision based 2D navigation

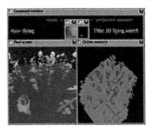

Figure 8.17:Vision based 3D navigation

Figure 8.18: Walking on sparse foothold locations

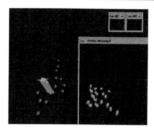

Figure 8.19: Walking on sparse foothold locations

Figure 8.20: Imported actor trajectories

Figure 8.21:Escaping from a maze

Figure 8.22:Walking through a flower field

Figure 8.23:Navigating humanoids

Figure 8.24: Actors of a tennis game

Figure 8.25:Tennis game

Figure 8.27:VLNET racket picking

Figure 8.26:VLNET tennis stade

Figure 8.28:Autonomous referee and player

Figure 8.29:VLNET tennis game

Figure 8.30:Force field - tree interaction

Figure 8.31:Future animation with humanoids

INDEX

Mathematics in Science and Engineering
Edited by C.K. Chui, Stanford University

Printed and bound by CPI Group (UK) Ltd, Croydon, CR0 4YY

08/05/2025

01864806-0001